SOCIAL PLAY IN PRIMATES

Academic Press Rapid Manuscript Reproduction

Proceedings of a symposium
concerning Social Play in Primates
held at the 1977 annual meeting of the
Animal Behavior Society at
Pennsylvania State University

SOCIAL PLAY IN PRIMATES

EDITED BY

EUCLID O. SMITH

Yerkes Regional Primate Research Center
and Department of Anthropology
Emory University
Atlanta, Georgia

ACADEMIC PRESS New York San Francisco London 1978
A Subsidiary of Harcourt Brace Jovanovich, Publishers

ACADEMIC PRESS, INC.
111 Fifth Avenue, New York, New York 10003

United Kingdom Edition published by
ACADEMIC PRESS, INC. (LONDON) LTD.
24/28 Oval Road, London NW1   7DX

Library of Congress Cataloging in Publication Data

Animal Behavior Society.
    Social play in primates.

    Contains papers presented at the 1977 annual meeting
of the Animal Behavior Society at Pennsylvania State
University.
    1. Primates—Behavior—Congresses.   2. Play behavior
in animals—Congresses.   3. Social behavior in animals—
Congresses.   I. Smith, Euclid O.   II. Title.
QL737.P9A57   1978   599'.8'045   78-14412
ISBN 0-12-652750-4

PRINTED IN THE UNITED STATES OF AMERICA

CONTENTS

CONTRIBUTORS

Numbers in parentheses indicate the pages on which authors' contributions begin.

Janice I. Baldwin (231), Department of Sociology, University of California, Santa Barbara, Santa Barbara, California

John D. Baldwin (231), Department of Sociology, University of California, Santa Barbara, Santa Barbara, California

Anna Bellisari (143), Department of Anthropology, The Ohio State University, Columbus, Ohio

Claud A. Bramblett (33), Department of Anthropology, University of Texas, Austin, Texas

J. A. Breuggeman (169), Department of Sociology and Anthropology, Purdue University, West Lafayette, Indiana

Douglas K. Candland (259), Program in Animal Behavior, Bucknell University, Lewisburg, Pennsylvania

Clinton Conaway (49), Department of Sociology and Anthropology, University of Rhode Island, Kingston, Rhode Island

Martin D. Fraser (79), Department of Mathematics, Georgia State University, Atlanta, Georgia

Jeffrey A. French (259), Program in Animal Behavior, Bucknell University, Lewisburg, Pennsylvania

Linda Haines (143), Department of Anthropology, The Ohio State University, Columbus, Ohio

Karen H. Hamer (297), Synanon Research Institute, Marshall, California

Carl N. Johnson (259), Department of Child Development and Child Care, University of Pittsburgh, Pittsburgh, Pennsylvania

Geoffrey Keifer (49), Department of Sociology and Anthropology, University of Rhode Island, Kingston, Rhode Island

James Loy (49), Department of Sociology and Anthropology, University of Rhode Island, Kingston, Rhode Island

Kent Loy (49), Department of Sociology and Anthropology, University of Rhode Island, Kingston, Rhode Island

Terry Maple (113), Department of Psychology, Emory University, Atlanta, Georgia

Elizabeth Missakian (297), Synanon Research Institute, Marshall, California

Donald Patterson (49), Department of Sociology and Anthropology, University of Rhode Island, Kingston, Rhode Island

Frank E. Poirier (143), Department of Anthropology, The Ohio State University, Columbus, Ohio

Euclid O. Smith (1, 79), Yerkes Regional Primate Research Center and Department of Anthropology, Emory University, Atlanta, Georgia

Donald Symons (193), Department of Anthropology, University of California, Santa Barbara, Santa Barbara, California

Evan L. Zucker (113), Department of Psychology, Emory University, Atlanta, Georgia

PREFACE

This volume is intended to look at what may be the most misunderstood and potentially confusing aspect of primate behavior. Interest in the study of play in primates has waxed and waned over the last decade as researchers have sought to describe it, define it, and determine its causation and evolutionary significance. Clearly, during this period many ideas have been accidentally rediscovered, quite innocently by many students of the problem. Unfortunately, in some ways we are little closer to the answers to the questions than we were a decade ago.

It is well documented that play is a particularly time-consuming behavior for the young. If we are to believe in the principles of Darwinian evolution and the applicability of evolutionary principles of behavior, then we must view the expression or potential for expression of play among primates to be under evolutionary control, to some extent. It seems then that if our goal is to explain behavior, then one must address the question of play in primates from this evolutionary perspective.

The authors of the papers in this volume participated in a symposium concerning "Social Play in Primates", held at the 1977 annual meeting of the Animal Behavior Society at Pennsylvania State University. Out of the papers and discussions at that meeting came a feeling that if play is to be understood within the Order Primates, then a variety of perspectives must be martialed to get at the heart of the problem. The papers represented here obviously reflect the particular biases of each individual author, but more importantly, they show the wide array of approaches toward understanding this most complex behavioral phenomenon, play.

In the opening chapter, I have attempted to set the stage for the volume by taking a historical look at the study of play. By tracing the development of an interest in or awareness of play in the young from the writings of Aristotle, through the educational reformers of the seventeenth and eighteenth centuries, to the psychologists of the nineteenth and early twentieth centuries, one gains an understanding of the bases of many of the various perspectives and theoretical orientations expressed in the literature today. It seems that, to some extent, researchers today concern themselves with either the structure of play, that is, the description and quantification of the behaviors involved in play, or its function in the behavioral repertoire of the individual as a member of a social group. It is out of these differences in perspective that the papers in this volume arise.

Bramblett makes a significant contribution to our understanding of play by demonstrating the importance of systematically collecting longitudinal data in such a manner that one can describe the ontogeny of play behavior. Furthermore, Bramblett's statistical approach to the analysis of longitudinal data should alert us all to the usefulness of regression analysis in describing data in a way that will allow useful inter and intraspecific comparisons. By the use of this statistical technique, Bramblett describes the gender differences in the acquisition and expression of play among vervets, and offers a useful comparison for other researchers.

Loy and his colleagues present important data on the effects of gonadectomy on social play in group-living rhesus monkeys. Their longitudinal study, lasting nearly four years, clearly demonstrates the powerful effect of gonadal hormones on the expression of play behavior. This study suggests that postnatal production of adult levels of testosterone is a key factor in the attainment of physical and behavioral maturity among males. The relationship between hormone levels and behavior is, admittedly, still unclear, but this study goes a long way in elucidating the mechanisms involved, particularly in regard to the expression of play behavior.

Attempts to define play among rhesus macaques and a review of the possible statistical techniques one might employ to aid in this process are the primary foci of my joint paper with Fraser. We have tried to develop a technique whereby the subjects of our investigation, in a real sense, dictate the classes of behavior that are scored. We have worked toward a structural definition of social play among rhesus macaques that can be tested against other species. The awareness and understanding of the complexity of play can best be achieved through quantification and systematic analysis in various primate taxa.

The great apes have held tremendous appeal for behavioral researchers for some time. Maple and Zucker make a unique contribution toward understanding play behavior by observations of three of the four great ape species. This comparative study is important, for it emphasizes the power of the ethological approach in understanding the nature and complexity of social behavior, particularly play. These observations show marked differences between the style of play among the species studied. The reader should be impressed with the similarities and differences in the play of our closely related nonhuman primate relatives.

The following three papers, those by Poirier *et al.*, Breuggeman, and Symons, offer a view of the function of play in primate society. Poirier and his colleagues review and organize much of what has been written on the function of primate play and correctly note that few tests of any of these functional hypotheses have been conducted. They conclude that the time has come to stop speculating and start collecting systematic data. Furthermore, they offer a series of testable hypotheses that could go far in clarifying the function of primate play.

Breuggeman presents an interesting perspective on the study of primate play by focusing, not on the young, but on the adults in a free-ranging rhesus group on Cayo Santiago. Interestingly, unlike the young, where play may facilitate learning and/or socialization, play among adult animals may be seen as a kind of social manipulation. She emphasizes this point by carefully noting the context for play among adult rhesus. Play among adults is a relatively infrequent event, but when it occurs, it can be quite dramatic. This paper certainly begins to fill the gap in our knowledge about play among adult nonhuman primates.

Symons makes an important theoretical statement differentiating play from aggression in young rhesus macaques, and concludes that the popular assumption that play serves to establish a dominance order untenable. More importantly, he brings the question of function to the fore by looking at the study of function in biology as compared to the social sciences. Many will disagree with his conclusions, but without a doubt, this paper will give sociobiologists as well as nonsociobiologists some food for thought.

Next, the Baldwins attempt to look at the ontogeny of play and exploration as the outcome of critical reinforcers on social behavior. They argue that sensory stimulation plays a much larger part than many have felt in the development of play and exploration and, in fact, set out a rather lengthy series of experiments that could be performed to explore the suggested relationships. Finally, the Baldwins make an important leap in the study of play and extend their data to human behavior. A careful reading of the last section of their paper offers some important implications for play in our own species.

From a theoretical point of view, the chapter by Candland and his associates makes an important contribution by developing a model for the study of play. This model offers an interesting perspective on categorization of various types of playful activities. The model is shown to have heuristic properties as well, for it allows for a reasonable method of categorizing the variability in play across the primate order. It is important, however, to note that they go beyond model construction and test some of their propositions. This paper is extremely useful in that it offers a new conceptual starting point for the study of play.

Finally, Hamer and Missakian examine play in young humans. They offer some significant conclusions on the nature of peer- vs. adult-oriented early social experience. From the point of view of our own social development, their chapter is one that is of considerable interest and offers interesting conclusions on sex differences in young of our species.

Overall, these chapters deal with a wide variety of perspectives, both theoretical and methodological, on the study of play. It is important that we continue to develop new and innovative ways of attacking the problem of understanding play in both human and nonhuman primates, lest we become complacent and lower our standards of intellectual comfort. As eclectic as these chapters may be, they should offer some additional insight into the mechanisms of play across the Order Primates.

In summary, I would be remiss if I did not express my gratitude to the people who have helped make this volume possible. Drs. Duane Rumbaugh and Sue Savage-Rumbaugh have stimulated my interest in play and offered support for the initial idea for a symposium on this topic. To my colleagues at the Yerkes Regional Primate Research Center, I offer my thanks for the many critical and insightful discussions over the material presented herein. Special thanks to Ms. Peggy Plant, who has assisted not only in preparation of the manuscripts, but offered her expertise in the coordination of all the material, as well as her unfailing good humor. Finally, thanks to my wife Trish, who may now have heard the last of play—for a while.

The preparation of this volume was supported in part by Grant RR-00165 from the U.S. Public Health Service, Division of Research Resources, National Institutes of Health.

A HISTORICAL VIEW ON THE STUDY OF PLAY
STATEMENT OF THE PROBLEM[1]

Euclid O. Smith

Yerkes Regional Primate Research Center and
Department of Sociology and Anthropology
Emory University
Atlanta, Georgia

I. INTRODUCTION

The study of play in animals has, historically, suffered from a variety of theoretical problems and a lack of empirical verification. "In 1945, Beach stated: 'Present-day understanding of animal play is regrettably limited and current views on the subject are considerably confused.' This could have been written today. Very little progress has been made in the past 25 years" (Müller-Schwarze, 1971:246). This is, perhaps, an overly severe indictment of the study of play in both human and nonhuman animals. This chapter attempts to set out the parameters of the problems manifest in the study of play by using a historical approach to the problem. It is quite clear that until those who are engaged in research on play set some reasonable standards of intellectual comfort, the study of play will continue to receive the criticism it has so justly deserved in the past from both biological and social scientists. If we intend to understand this complex behavioral event called play, then a clear perception of the problem is required.

At present, we are experiencing a growing interest in the study of play. Anthropologists, ethologists, psychologists, sociologists, among others, are becoming aware that the study of play is an important area of interest if we are to under-

[1]*The preparation of this paper was supported by Grant RR00165, U. S. Public Health Service.*

stand behavioral development and social organization among
animals (Bekoff, 1973a). This interest is documented by the
increasingly large number of studies of play in a variety of
animals.[2]

Since publication of Beach's (1945) classic work, there
have been a number of theoretical reviews of play (Baldwin and
Baldwin, 1977; Bekoff, 1972, 1976a; Berlyne, 1969; Dolhinow
and Bishop, 1970; Fagen, 1974, 1976; Gilmore, 1966; Herron and
Sutton-Smith, 1971; Hutt, 1970; Loizos, 1966, 1967; Lorenz,
1956; Meyer-Holzapfel, 1956; Millar, 1968; Miller, 1973; Müller-Schwarze, 1971; Poirier and Smith, 1974; Welker, 1961,
1971). However, a concise definition of play has not emerged
from these reviews, nor has a widely accepted view of the function of play. Darling noted (1937) that play is a phenomenon
easier to describe than explain, while Hurlock (1934), on the
other hand, stated that there is little concern over the definition of play among writers. The importance of play in normal social development has been suggested by a number of

[2]*For example, Brownlee (1954) on domestic cattle* (Bos tau-
rus), *Chepko (1971) on goats* (Capra bircus), *Müller-Schwarze
(1968) on black-tailed deer* (Odocoileus hemonius columbianus),
Poole (1966) on polecats (Putorius putorius putorius), *Poole
and Fish (1975) on rats and mice* (Rattus norvegicus *and* Mus
musculus), *Wilson and Kleiman (1974) on three cavimorph rodents* (Octodon degus, Octodontomys gliroides *and* Pediolagus
salinicola), *Rensch and Dücker (1959) on mongooses* (Herpestes
ichneumon L.), *Schenckel (1966) on lions* (Panthera leo),
Steiner (1971) on Columbian ground squirrels (Spermophilus columbianus columbianus), *Tenbrock (1960) on red foxes* (Alopex
lagopus), *Wilson (1973) on voles* (Microtus agrestis), *Wilson
(1974) and Wilson and Kleiman (1974) on seals* (Phoca vitulina
vitulina, Halichoerus grypus *and* Phoca vitulina concolor),
Wilson and Kleiman (1974) on pygmy hippopotami (Choeropsis li-
beriensis) *and giant pandas* (Ailuropoda melanoleuca), *Farentinos (1971) and Gentry (1974) on sea lions* (Eumetopias jubata),
Bekoff (1974) on canids (Canis sp.), *Henry and Herrero (1974)
on black bears* (Ursus americanus), *Lazar and Beckhorn (1974)
on ferrets* (Mustela putorius), *Welker (1959) on raccoons* (Pry-
ocon lotor), *Wemmer and Fleming (1974) on meerkats* (Suricata
suricatta), *Baldwin and Baldwin (1973, 1974) on squirrel monkeys* (Saimiri sp.), *Fedigan (1972) on vervets* (Cercopithecus
aethiops), *Dolan (1976) on Sykes' monkeys* (Cercopithecus mitis
kolbi), *Symons (1973), Redican and Mitchell (1974), Lichstein
(1973a,b), Meier and Devanney (1974), Smith (1977), et al., on
rhesus macaques* (Macaca mulatta), *Welker (1954, 1956a,b), Bierens de Haan (1952), et al., on chimpanzees* (Pan troglodytes),
and Freeman and Alcock (1973) for gorillas (Gorilla gorilla
gorilla) *and orang-utans* (Pongo pygmaeus pygmaeus).

researchers; however, satisfactory definitions and empirically tested theoretical propositions about play are scarce.

A literature review reveals considerable variability in what has been called play. Hutt (1966) noted that play included such widely divergent behavior patterns as the darts and gambols of young birds and mammals to the ritualized games of adult humans. Clearly, this suggests that the assumption has been made that play in all species arises from similar motivational sources (Meyer-Holzapfel, 1956). The extension of the use of the term play in its colloquial sense from human to nonhuman animals further confounds the issue, and has contributed to the confusion surrounding this class of behavior.

Historically, play as a behavioral category was first mentioned by Plato (Millar, 1968), who recognized its practical value in the development of young individuals. Aristotle, too, thought children should be encouraged to play at what they were to do seriously as adults. Educational reformers, from Comenius in the seventeenth century to Rousseau and Pertalozzi in the eighteenth and early nineteenth centuries, accepted that education should consider all aspects of child development, including play. These general philosophical ideas culminated in Frobel's stress on the importance of play in learning (Millar, 1968), but were of limited theoretical value. However, as early as the mid-nineteenth century, a number of theoretical views of play were beginning to emerge in the literature.

Regardless of the definitional problems associated with play and confusion over its causation and function, play is restricted to homiotherms, with a few exceptions among the poikilotherms [see Fagen (1976) for a listing of the poikilotherms that have been observed to engage in playful behavior, as well as a suggested evolutionary basis for this difference]. Welker (1961) has developed an exhaustive list of different taxa which have been reported to play.

II. THEORIES OF PLAY

It is important to consider the variety of theories of play to appreciate the problems surrounding this behavior. Out of a plethora of theories have come attempts at definitions and descriptions of play in a wide array of species, which must be considered in an evolutionary perspective to understand the diversity of expression of play behavior.

Theories of play can be broadly classed into two general categories: 1) those dealing with the developmental aspects of play in human and nonhuman mammals; and 2) those dealing with human imaginative play. Since some have assumed that all play arises from the same motivational state (Meyer-Holzapfel, 1956), has the same physiological base, and often is identical, or similar in form (Rensch, 1973), a brief review of both types of theories is presented.

A.   *Developmental Aspects of Play*

1.   Surplus Energy Theory. One of the oldest theoretical statements concerning the significance of play is attributed to Schiller (1875) and Spencer (1873), although it may have had its origin in the writings of seventeenth and eighteenth century educational reformers. Schiller (1875) called play the expression of exuberant energy. Curti (1930) noted that Schiller merely suggested that play occurred when an ample supply of energy was available, although later writers offered considerable reinterpretation of the original statement. Spencer (1873) elaborated on Schiller's notions by suggesting a neurophysiological basis for this "excess energy". According to Spencer (1873), nerve centers disintegrate with use and need time to be restored. A nerve center which has been at rest for a considerable period becomes physically unstable and, therefore, is over-ready to respond to any kind of stimulation. This instability of neural centers was interpreted as surplus energy; hence, the common name of the general theory.

Briefly, this position holds that play results from surplus energy which exists because the young are freed from self-preservation through parental action. The surplus energy theory postulates a quantity of excess energy available to the organism, and a tendency to expend this energy, even though it is not necessary for the maintenance of life (Gilmore, 1966). The surplus energy theory has enjoyed widespread appeal and has been presented in a variety of other forms (Alexander, 1958; Tinklepaugh, 1934; Tolman, 1932).

Beach (1945) notes the following objections to the surplus energy theory: 1) These notions are based on circular reasoning. The catch lies in the definition of "surplus" - the decision as to whether or not expended energy is surplus depends on the interpretation of the behavior as playful or "serious". 2) "In the area of mental or emotional energy...it is sheer nonsense to predicate explanations of behavior upon supposed accumulation and discharge of hypothetical forces. Definition of one unknown in terms of a second unknown is good algebra, but poor psychology" (Beach, 1945:528).

Lorenz's (1950) psychohydraulic model of motivation is consistent with the surplus energy theory; however, both falter mainly in the fact that there is little evidence that physical energy can be stored in an organism like water in a reservoir (Bekoff, 1973a). Mitchell (1912) objected to the surplus energy theory based on the interpretation that the discharge of energy was a "waste product". Morgan (1900) also noted that "normal" rather than "surplus" energy is involved in the play of animals. Groos (1898), among others, has observed that superabundant energy is not always a condition of play. Animals will play to apparent exhaustion and be ready to play again with a very brief rest or no rest at all (Beach, 1945). Hinton and Dunn (1967) note that the unfortunate shortcoming of the surplus energy theory is that it directs attention away from the selective dimensions of play and implies that play is its own motivation.

Others have developed a variant on this theory, noting that play is the result of an inner drive, is intrinsically rewarding, or done for sheer pleasure (Aldis, 1975; Buhler, 1930; Dobzhansky, 1962; Döhl and Podolczak, 1973; Dolhinow and Bishop, 1970; Eibl-Eibesfeldt, 1975; Gehlen, 1940; Lazarus, 1883; Morris, 1962; Patrick, 1916). Perhaps the only support for this theory lies in the fact that young animals play more than do adults (Cooper, 1942); however, the reasons for this seem to be much more complex than those implied in this theory (Bekoff, 1973a).

2. *Relaxation Theory.* Another classical theory views play not as a product of surplus energy, but resulting from a deficit of energy. Lazarus (1883) and Patrick (1916) noted that play is a mode of dissipating inhibition resulting from fatigue due to relatively new tasks to the organism. Therefore, play most frequently occurs in the early developmental stages and replenishes energy for the unfamiliar cognitive activities of the young (Gilmore, 1966). Winch (1906a,b) added a neurological interpretation of the relaxation theory when he noted that play exerted little demand on the higher nervous centers.

3. *Optimal Arousal Theory.* Baldwin and Baldwin (1977) note that sensory stimulation which serves to keep the individual within an optimal arousal zone is reinforcing; but overstimulation or understimulation is aversive. The tendency for young animals to seek an optimal arousal level has been well documented for nonhuman primates in both field and laboratory studies (Harlow and Harlow, 1965, 1969; Jay, 1965; Mason, 1965, 1967, 1968, 1971; Schaller, 1963; van Lawick-Goodall, 1967). From clinging to an arousal-reducing mother to arousal-increasing play and exploration, the young nonhuman primate vascillates during much of the waking day. Andrew (1974), Berlyne (1969), Bindra (1959), Hebb (1949, 1955), Leuba (1955), Schneirla (1959), and Welker (1956a,b), among others, have elaborated the arousal theory in various fashions. Finally,

Heckhausen (1964) notes that play may be the behavioral event
that strives to keep neural activity at an optimum level.

    4. Pre-exercise Theory. Karl Groos (1898, 1908) presen-
ted a theory of play based broadly on natural selection, in
which he emphasized that only animals best fitted to cope with
the environment survive. If animals play, it is because play
is useful in practicing skills needed in later life. Only
animals endowed with detailed instinctive patterns which are
perfect on the first trial have no need to play (Millar, 1968).
Consequently, some animals must practice and perfect their incomplete hereditary skills before a serious need to exercise
them arises (Poirier and Smith, 1974). Groos (1898) draws
heavily on accounts of play-fighting among young animals as
support for his pre-exercise theory.
    Pycraft (1913) claims that Groos' theory makes infancy
seem an irresponsible apprenticeship to the seriousness of
life. Millar (1968) notes that the play of mature animals is
less easily accommodated by a theory that play is an instinct
to practice instincts used in adult life.
    These "practice" hypotheses presuppose that all social
learning is adaptive; however, they only partially explain
playful behavior (Fagen, 1974). Clearly, those who differentiate between "experimental" practice (play) and simple practice (Beach, 1945; Bruner, 1973a,b) imply that the special
structure of play is an observable correlate of a "playful"
learning mechanism. These observers indicate that playful
practice requires varied experiences, and stress the interactions of skill and the environment, while rote practice per-
fects the application of a specific behavior pattern (i.e.,
infant transport) without contributing behavioral flexibility
(Fagen, 1974). Similarly, Loizos (1967) states that play is
not necessary for practice of adult behaviors.
    Bekoff (1973a) notes that Gross' theory ignores the social
importance of play for the developing organism. Furthermore,
recent research (Fox, 1969; Poole, 1966) has demonstrated that
many instincts required for "serious" adult life tend to be
unmodified by early experience. Nonetheless, Groos' pre-exercise theory underscores one important dimension of play, the
necessity of exercising various motor patterns (coordination
of reflexes, muscular and skeletal development) (e.g., Brownlee, 1954; Fagen, 1976), but ignores the impact of this highly
social behavior on the developing organism (Bekoff, 1973a).

    5. Recapitulation Theory. In 1906, G. Stanley Hall posi-
ted the recapitulation theory of play. This theory rests on
the notion that children are a link in the evolutionary chain
between human and nonhuman animals, and pass through all the
stages from protozoan to human in their lives as embryos

(ontogeny recapitulates phylogeny). Hall (1906) extended the notion of recapitulation to the whole of childhood, and claimed that the child passes through a series of play stages corresponding to, and recapitulating, the cultural stages in the development of races (Gilmore, 1966). Winch (1906a,b) notes that the recapitulation theory, simply stated, says the work of the father becomes the play of the children, and the past holds the key to all play activities (Lehman and Witty, 1927). Admittedly Lamarckian in perspective, Hall's recapitulation theory served to stimulate interest in the developmental behavior of children (Millar, 1968).

   6. *Growth Theory*. Appleton (1910) suggested another position with respect to play. She concluded that play is a response to a generalized drive for growth in the organism, although not instinctual pre-exercise as envisioned by Groos (1898, 1908). A child plays because it "knows" that play is the method by which it will grow (Gilmore, 1966). This theory seems to differ little from Groos', except the organism is supposedly conscious of its activities, and is exercising its drive for growth (Bekoff, 1973a). Many of the same criticisms initially directed toward the pre-exercise theory of Groos can be leveled against Appleton's growth theory.

B. *Human Imaginative Play*

   1. *Infantile Dynamic Theories*. In the late nineteenth and early twentieth centuries, theories of play were developed which differed from classical theories primarily in that they invoked explanations based on dynamic factors of individual personality, and were designed to explain individual variability in play behavior (Gilmore, 1966). Buytendijk (1934) rejected earlier pre-exercise theories and interpreted play, which for him was directly associated with object manipulation, as a product of youthful dynamics (Meyer-Holzapfel, 1956). Buytendijk (1934) states that a child plays because it is a child. Gulick (1920) notes that if you want to know what a child is, study its play; if you want to effect what the child will become as an adult, direct its play (Ghosh, 1935).
   Lewin (1933) suggests that play occurs because the individual's cognitive life space is unstructured, resulting in a failure to discriminate between the real and unreal. In general, the infantile dynamics theories rest on the proposition that play is the child's way of thinking (Gilmore, 1966).
   Of all infantile dynamics theorists, Piaget (1951) is perhaps the best known. Generally, Piaget viewed play as the product of a stage of intellectual development, through which the child must pass in developing from the original egocentric

and phenomenalistic viewpoint to the adult objective and ra-
tionalistic outlook (Gilmore, 1966). Play, for Piaget, seems
to be a process whereby the individual fits bits of information into an existing conceptual schema. Additionally, Piaget's theory makes a sharp distinction between causes and
effects of play. His entire theoretical orientation towards
play is intimately related to his general theories of cognitive development. Piaget's work, however, is closely related
to cathartic and psychoanalytic theories of play (see below).

   2. *Cathartic and Psychoanalytic Theories.* Mitchell and
Mason (1934) suggest that Aristotle may have offered the earliest thoughts on the cathartic theory of play. Generally,
the cathartic theory views play as the child's attempts to
master situations which, at first, were too difficult. The
cathartic theory was first suggested by Carr (1902); however,
others were soon to follow with variants of the main theme
(e.g., Curti, 1930; Reaney, 1916; Robinson, 1920).
   The psychoanalytic theory of play is the most recent variation of the general cathartic theories. Freud (1955, 1959a,b)
developed the psychoanalytic theory of play which is, of
course, only a small portion of his more general theory of
psycho-social development. Play, for Freud, shares many of
the same unconscious components which shape dream life, and in
this sense is somewhat similar to Piaget's notion of play.
Freud conceived of play as closely related to fantasy; in
fact, he defined play as fantasy woven around real objects, as
contrasted to daydreaming which is pure fantasy (Gilmore,
1966). Erikson (1950, 1951) made additional contributions to
the psychoanalytic theory of play by emphasizing the coping
and anxiety reducing aspects of play.
   These psychoanalytic theories of play are theoretically
sophisticated and well developed; however, they have led to
little empirical research (Gilmore, 1966). Gilmore (1964)
seems to have made one of the few attempts to experimentally
test these theories of play. In general, psychoanalytic theories have dealt with the behavior of children in terms of
adult behavior and have not been concerned with the percep-
tions of the child (Bekoff, 1973a).
   In sum, these theories of play have led to confusion rather than clarity in explaining and understanding play behavior. To some extent, this confusion rests in the colloquial
use of the term and its interchangeability from human to nonhuman mammals. Clearly, there is an intuitively recognized
similarity between the play of children and young animals
(Müller-Schwarze, 1971) which adds to the confusion, although
it is precisely because of these similarities that there is
high interobserver agreement on when animals are playing
(Bekoff, 1973a,b; Loizos, 1966, 1967; Miller, 1973). Some

suggest (Berlyne, 1960, 1969; Müller-Schwarze, 1971; Schlosberg, 1947; Welker, 1971) that the generic term, play, has become so confused that it should be abandoned in favor of more precise terminology, i.e., ludic behavior, motor play, exploratory play.

Additionally, some of the confusion over a theoretical basis for play lies in a failure to understand the conceptual difference between theories of the causation of play and theories of the function of play. For example, the surplus energy theory, the relaxation theory, and the optimal arousal theory are concerned with the underlying causal basis for play; while the pre-exercise theory is a theory of the function of play. It may be that these various theories have invoked different levels of explanation to account for the same phenomenon.

III.   FUNCTIONALISTS VS. STRUCTURALISTS -- DIFFERENT VIEWS OF
       THE SAME PHENOMENA

Fagen (1974) has clarified much of the theoretical literature by suggesting that there are two vastly different conceptual positions with respect to the study of play (e.g., Gilmore, 1966; Sutton-Smith, 1971). "Functionalists study the causes of play, including underlying behavioral mechanisms and/or possible adaptive significance (e.g., Ewer, 1968), whereas structuralists consider the form and appearance of play (e.g., Loizos, 1966, 1967; Müller-Schwarze, 1971)" (Fagen, 1974:851). Playful behavior for the functionalist is necessarily play at something [e.g., play mothering (Lancaster, 1971), play fighting (Aldis, 1975; Symons, 1973)], while the structuralist claims play has a particular unique structure (Fagen, 1974).

A. *Structuralists*

This position concerns itself with careful description of the behavior itself, and manifests itself in a variety of descriptions and definitions, although some researchers have noted that because there is a consensus on what is play, a precise definition is not required (Lorenz, 1956; Thorpe, 1956). Welker (1971) notes that play and, for that matter, exploration are not unique behavioral categories distinct from other elements of the behavioral repertoire. Furthermore, precise differentiation of any behavioral category on any but a neurophysiological basis seems unfruitful according to Welker (1970), although some would disagree.

1. Categories of Play. These differences not withstand-
ing, Millar (1968) has suggested four general classes of play:
1) general activity where there is not some immediate response
to an environmental stimulus (e.g., gambolling of puppies,
frolocking of lambs); 2) parts of behavior patterns which normally lead to fulfillment of a definite biological function,
but occur out of context, or without accomplishing the purpose
(e.g., play-fighting, sex play of sexually immature animals,
etc.); 3) interactions involving at least two animals and occurring mainly between members of organized groups, and may
overlap considerably with (2) (e.g., parental play, play-fighting, etc.); and 4) activities which include investigating and
manipulating the environment and experimenting with objects
(Kohler's chimpanzees, Menzel's chimp's invention of ladders,
etc.)

Müller-Schwarze(1971:230) notes, "Despite the confused terminology and difficulties of new definitions for the various
kinds of 'play', there exists one distinct category, motor
play. These are juvenile social and solitary behaviors which,
in context, sequence and function, differ clearly from other
behaviors." Mears and Harlow (1975) have offered a further
classification of play, self-motion play or peragration, which
is defined as motion of the self as a reinforcer. They further
suggest that self-motion play can be either social or nonsocial.

2. Theoretical Definitions of Play. Clearly, many researchers have asserted that play exists, but relatively few
have attempted precise, unambiguous definition. Although Millar (1968) has established categorical distinctions of types
of play, she never precisely defines play. However, some definitions of play have been presented in the literature: 1) Bekoff (1972:417) has stated, "Social play is that behavior which
is performed during social interactions in which there is a
decrease in social distance between the interactants, and no
evidence of social investigation or of agonistic or passive-
submissive behaviors on the part of the members of a dyad
(triad, etc.), although these actions may occur as derived acts
during play. In addition, there is a liability of the temporal
sequence of action patterns, actions from various motivational
contexts." 2) Fagen (1974:850) notes that play is "...active,
oriented behavior whose structure is highly variable, which
apparently lacks immediate purpose, and which is often accompanied by specific signal patterns." 3) Müller-Schwarze (1971:
223) provides a definition of motor play as "...the performance
of a mixed sequence of mostly stereotyped behavior patterns by
an immature animal. These patterns belong to different functional systems and do not serve their usual functions. The
patterns often occur in a social situation under moderate

general arousal, but low specific motivation." 4) In a study
of social play in free-living baboons, Owens (1975:387) notes
that play "...is generally composed of behavior patterns seen
in other functional contexts...but lack of immediate biological end is used here as a distinguishing criteria." Additionally, Owens (1975) describes the movement patterns in aggressive, sexual, and parental play, his three classifications of
social play. 5) Poole and Fish (1975:63), in a study of play
in rats and mice, define play as "...apparently goalless behavior in which the movements were energetic and exaggerated.
This was particularly noticeable if the play behavior pattern
were compared with the nearest equivalent form of adult behavior, performed in the usual context."

    3. Operational Definitions of Play. Problems with seman-
tics have caused considerable confusion and prompted a number
of researchers to abandon their attempts at theoretical definition. Many now favor description in operational terms of
precisely what behavior patterns are seen in playful interactions. For example: 1) Aldis (1975) notes that "...almost
everyone would agree that chasing and play-fighting in young
animals is play...these behaviors are usually accompanied by
play signals and are modified in certain ways from their serious counterparts." 2) Wilson (1974:38) describes play in
seals *(Phoca vitulina vitulina* and *Halichoerus grypus)* as
"...leaping and splashing in the water and exaggerated flap-
ping towards one another over the rocks or shore." 3) Baldwin
and Baldwin (1974:304) describe social play in squirrel monkeys *(Saimiri* sp.) as "...social interactions that include
wrestling, chasing, sham-biting, jumping on, pulling tails,
carrying, steep leaps, and other related activities." 4) Freeman and Alcock (1973), in a study of juvenile interactions in
gorillas *(Gorilla gorilla gorilla)* and orang-utans *(Pongo pygmaeus pygmaeus)* describe social play as including wrestling
and tug-of-war encounters. 5) Welker (1961:175-176) defines
play as "...a wide variety of vigorous and spirited activities:
those that move the organism or its parts through space such
as running, jumping, rolling, and somersaulting, pouncing upon
and chasing objects or other animals, wrestling, and vigorous
manipulation of body parts or objects in a variety of ways."
6) Harlow and Harlow (1965), in addition to discussing the
ontogeny of interactive play in young rhesus macaques *(Macaca
mulatta)*, describe three types of social play. Rough-and-tumble play consists of infant monkeys, rolling, wrestling and
sham-biting each other without injury and seldom becoming
frightened. Approach-withdrawal play, a more complex type,
involves mutual chasing in which physical contact is minimized.
"Genteel rough-and-tumble play, in the second year of life,
becomes superceded by a kind of physical contact and release

can be physically painful and the biting responses may evoke cries of distress and anguish from the monkey being bitten" (Harlow and Harlow, 1965:313).

4. *Structural Characteristics of Play.* Often observers simply characterize playful activities apart from non-playful ones. Beach (1945) suggests the following outstanding characteristics of play:

(a) Play is typically thought to have an emotional element of pleasure associated with it (e.g., Bekoff, 1974; Bertrand, 1969; Csikszentmihalyi and Bennett, 1971; Loizos, 1966; Poole, 1966). Washburn (1973:130) suggests that, "Judged by their behaviors, play is pleasurable to the young primate." Although we may be able to operationally define the pleasurable elements of play, how could this be tested?

(b) Play is generally thought to be characteristic of immature rather than adult animals; however, adults of many species do play, particularly in the mother/offspring context (Altman, 1966; Fox, 1971; Jay, 1963; Jolly, 1966; Kruuk, 1972; Rheingold, 1963). However, instances of adult play have been reported outside the mother-offspring network [e.g., free-ranging bonnet macaques, *Macaca radiata* (Simonds, 1965); captive squirrel monkeys, *Saimiri sciureus* (Winter, 1968); free-ranging mountain sheep, *Ovis canadensis* (Geist, 1971); captive (Bingham, 1927) and free-ranging chimpanzees, *Pan troglodytes* (van Lawick-Goodall, 1968); free-ranging hyenas, *Crocuta crocuta* (Kruuk, 1972); captive (Redican and Mitchell, 1974), enclosed (Gordon, Rose and Bernstein, 1976), and semi-free-ranging rhesus macaques, *Macaca mulatta* (Breuggeman, 1976; Kaufmann, 1967); and others].

(c) Play differs from non-playful responses in having no relatively immediate biological result or when benefits are delayed until a later age. In other words, play is customarily regarded as non-utilitarian, with no immediate purpose (Fagen, 1974).

(d) The expression of play is species specific. For a description and comparison of social play in two platyrrhine monkeys (*Saimiri sciureus* and *Allouatta palliata*), see Baldwin and Baldwin (1977). Wilson and Kleiman (1974) offer an excellent comparative study of play in three South American rodent species (*Octodon degus, Octodontomys gliroides* and *Pediolagus salinicola*).

(e) The amount, duration and diversity of play in a given species may be related to certain ecological characteristics. In general, primates and carnivores tend to play more than most other mammals (Baldwin and Baldwin, 1977), while play typically decreases in frequency as one descends the phylogenetic scale (Aldis, 1975; Eibl-Eibesfeldt, 1975; Welker, 1961). Meyer-Holzapfel (1956) adds the following:

(f) Play is an expression of a high general motivational state, not a specific motivation. Contrarily, Mason (1965) suggests play can be suppressed when young primates are overly stimulated, and play typically occurs in moderately arousing environments. Furthermore, Eibl-Eibesfeldt (1975:276) notes that "...there exists a specific motivation for play which is based on a curiosity drive; that is, a mechanism which moves the animal to seek new situations and to experiment with new objects."

(g) Behavioral elements which comprise play are drawn from a variety of contexts, and the elements are mixed and replaced by each other in an irregular manner (Baldwin and Baldwin, 1977; Bekoff, 1972; Dolhinow and Bishop, 1970; Loizos, 1966, 1967; Marler and Hamilton, 1966). However, Müller-Schwarze (1971) found sequential stability in sequences of play behavior in blacktailed deer (*Odocoileus hemonionus columbianus*).

(h) Depending on the motivational level, play can be released by non-specific external stimuli or by no discernible stimuli at all. Eibl-Eibesfeldt (1975) clarifies this point by noting that motivating mechanisms present in the "normal" expression of a behavior pattern are frequently absent in its playful expression. However, identification of discernible stimuli responsible for elicitation of play behavior may be difficult, as "natural contingencies of reinforcement" (Baldwin and Baldwin, 1977) obscure the traditional stimulus-response design.

(i) Play occupies a relatively low position in the hierarchy of types of behavior, and occurs only when the animal's essential needs have been met, and not in stressful situations (Poole and Fish, 1975). Several field reports have clarified the nature of the relationships between food ecologies and play, and pointed out the effect of abundance and dispersion on its expression. Baldwin and Baldwin (1972, 1973), Hall (1963) and Loy (1970) have documented the effect of food deprivation on the reduction in frequency of the expression of play in free-living primates. Rosenblum, Kaufman and Stynes (1969) for pigtail macaques (*Macaca nemestrina*), Southwick (1967) for rhesus macaques (*Macaca mulatta*) and Baldwin and Baldwin (1976) for squirrel monkeys (*Saimiri sciureus*) found a significant decrease in play with a reduction in food supply in controlled laboratory investigations. Poole and Fish (1975) suggest some additional characteristics:

(j) Play can be recognized by its exaggerated movements. Play sequences are often a collection of disrupted activities, extravagent, uneconomical, clumsy, and fragmented, unlike the orderly and efficient behavior of adults (Baldwin and Baldwin, 1977; Marler and Hamilton, 1966; Miller, 1973).

(k) Finally, play may be characterized as having certain exclusive behavior patterns which distinguish it from "serious"

behavior (Dolan, 1976). It has been suggested that these behavior patterns signal a readiness to play, and thereby communicate to potential interactants an individual's intent.
Altmann (1967) described communication as a process whereby
the behavior of one individual affects the probability of behavior of another; and further notes that the development of
a system of metacommunication, communication about communication, allows an individual's full participation in all adult
behaviors. In a study of coyotes (*Canis latrans*), Bekoff
(1975) found certain metacommunicative signals employed in social play bouts to establish a "play mood". Bateson (1955a,b)
noted that play can only occur when primates are capable of
some degree of metacommunication in that their reference is
the interaction (Bateson, 1955a,b).

The most frequent specific signal, the primate play face,
has been extensively reviewed (Altmann, 1962; Chevalier-Skolnikoff, 1973, 1974; Goodall, 1965; Loizos, 1967; van Hooff,
1967, 1972; van Lawick-Goodall, 1968). Sade (1973) notes that
transverse body rotation functions as a play specific invitation in rhesus macaques, while Struhsaker (1967) reports a
play call for vervet monkeys. Play vocalizations are known
for other primate species, e.g., squirrel monkeys, *Saimiri
sciureus* (Winter, Ploog and Latta, 1966), chimpanzees, *Pan
troglodytes* (van Lawick-Goodall, 1968), and gorillas, *Gorilla
gorilla beringei* (Schaller, 1963). For non-primate mammals,
play specific behaviors have also been observed in blacktailed
deer, *Odocoileus hemonionus columbianus* (Müller-Schwarze,
1971), canids, *Canis* sp. (Bekoff, 1973b), Columbian ground
squirrels, *Spermophilus columbianus columbianus* (Steiner,
1971), domestic cats, *Felis catus* (West, 1974) and seals,
Phoca vitulina vitulina and *Halichoerus grypus* (Wilson, 1974).

By identifying salient characteristics of play, we are
indeed closer to a definition; however, it should be noted
that there exist differences of opinion regarding the characteristics of this behavioral constellation (see b, f, g and i
for examples). Beach (1945:538) noted correctly that "...it
should be recognized that no single hypothesis can be formulated to explain all forms of play in every animal species. The
types of activity which are commonly termed playful are so
variable in form and complexity that a different interpretation is indicated, at least for each major category." Bekoff
(1974:228) has also stated that "...it is safe to conclude
that no one theory or explanation is applicable to all animals." Finally, Poole (1966) cautions that a study of play in
a single species does not allow formulation of a general theory of play.

Today, some researchers are not attempting a broad theoretical definition of play applicable to all species, but are
defining playful behavior for individual species in specific

settings. Furthermore, as has been discussed, play may be recognized by a number of characteristics, intrinsic to the behaviors themselves (e.g., exaggerated, uneconomical movements, etc.), but as Bateson (1955a,b) has noted, the contextual aspects of particular behaviors should not be overlooked in forming a definition. In other words, behaviors may be essentially similar in form and differentiated in meaning only by context (Breuggeman, 1976).

As with other categories of behavior (e.g., aggression, submission, caregiving), a definition of play must have as its basis a precise, unambiguous description of the motor patterns involved. However, identification of the precise motor patterns is only a first step. Typically, through an intuitive idea of what motor patterns constitute a particular class of behavior, researchers have lumped these molecular units of behavior into larger, functional molar classes (e.g., sex, aggression, etc.). From the high interobserver agreement on the identification of play as a behavioral phenomenon, one might infer that there are some interspecific characteristics of this behavior (Miller, 1973). However, often these intuitive behavioral typologies rely only on face validity. Frequently, these intuitive typologies are anthropomorphic and, to some extent, egocentric in perspective. Play is particularly susceptible to these abuses, for when one engages in any activity which does not fulfill primary needs, it is often characterized as play (Loizos, 1966, 1967). On the other hand, Reynolds (1967) suggests that play is really the "work" of immatures, as it is accompanied by intense concentration and a motivation for continuation until mastery is achieved.

To identify molar behavioral classes without ascribing underlying motivation is a difficult task. Only in recent years have researchers begun to be concerned that, through the labels we attach to particular behaviors, a consistency in animal behavior is being projected which may or may not be present in reality (Baldwin and Baldwin, 1977). The quest for a definition of play, clearly a molar class of behavior, in some respects is representative of definitional problems with other behavioral categories (e.g., aggression). However, factors intervene in a discussion of play which are not so apparent in other behavioral classes, namely the cross-specific, anthropomorphic recognition of playful behaviors. Marler and Hamilton (1966) noted that play is a troublesome class of behavior to define; however, the difficulties should not obviate any attempts at precise definition.

B. *Functionalists*

Instead of concern over structural or definitional problems, functionalists are concerned with the adaptive significance of play for the organism. Müller-Schwarze (1971:240)
notes that, "The amount of time and paper spent on speculations on possible functions of motor play in immature animals
is in inverse proportion to the amount of facts available on
the question." Certainly, numerous functions for play have
been postulated, although many without adequate data (Beach,
1945; Welker, 1971). Such speculations have, in some instances, had the effect of generating testable hypotheses. It
should be noted (Baldwin and Baldwin, 1977) that there are alternative kinds of experiences that may produce the same functional ends as play and exploration (i.e., normal social development) and that play can result in dysfunctional (maladaptive) consequences. Many researchers have argued, quite
convincingly, that play is adaptive and functional, both for
the individual and the species (e.g., Bekoff, 1972; Dolhinow
and Bishop, 1970; Loizos, 1967; Poirier, 1969, 1970, 1972;
Rensch, 1973; Suomi and Harlow, 1971; Washburn, 1973).

1. Functions of Play. Numerous functions have been
ascribed to play [for a list of 30, see Baldwin and Baldwin
(1977)], but they can be organized into five general categories:

a. Physical development. Many researchers have suggested that play offers an opportunity for physical stimulation
necessary for proper development of muscle tissue, skeleton,
and the central nervous system, as well as developing motor
skills essential for survival (Beach, 1945; Brownlee, 1954;
Dobzhansky, 1962; Dolhinow and Bishop, 1970; Ewer, 1968;
Fagen, 1976; Groos, 1898; Hinde, 1971; Levitsky and Barnes,
1972; Poirier, 1970; Riesen, 1961, 1965; Southwick, Beg and
Siddiqi, 1965; Volkman and Greenough, 1972; West, 1974).

b. Social development. Many researchers note that play
is an important aspect of normal psycho-social development
(Baldwin, 1969; Dolhinow, 1971; Dolhinow and Bishop, 1970;
Eibl-Eibesfeldt, 1967; Fedigan, 1972; Harlow and Harlow, 1965,
1969; Miller, 1973; Poirier, 1972; Poirier and Smith, 1974;
Welker, 1961; White, 1959). Furthermore, play may allow the
developing individual to gain valuable information about the
environment (Baldwin, 1969; Birch, 1945; Eibl-Eibesfeldt,
1967; Fedigan, 1972; Lancaster, 1971; Loizos, 1967; Lorenz,
1956; Schiller, 1957; Symons, 1973; Tsumori, 1967; Washburn
and Hamburg, 1965). Historically, some theorists felt that
play actually served as practice for adult activities (Groos,

1908; Mitchell, 1912; Pycraft, 1913), a position which in recent years, has received renewed attention (Dolhinow and
Bishop, 1970; Hansen, 1966; Suomi and Harlow, 1971; Washburn
and Hamburg, 1965). Loizos (1967) points out, however, that
it is not necessary to play in order to practice, and as Poole
(1966) noted, play in polecats is stereotyped and unmodified
by experience.

     c. Establishment of dominance hierarchy. Carpenter
(1934) has suggested that play may facilitate learning a position in the social order. Although dominance among juveniles
may be partially a function of maternal rank (Koford, 1965;
Loy and Loy, 1974; Marsden, 1968; Sade, 1967), relative size
(Symons, 1973), or seniority in the group (Drickamer and Vessey, 1973; Vessey, 1971), during play juveniles gain experience and become familiar with dominant and subordinate situations (Dolhinow and Bishop, 1970; Hall and DeVore, 1965; Harlow and Harlow, 1965; Jay, 1965).[3] Rhine (1973) calls social
play "behavior testing", where active social experimentation
allows individuals to determine each other's strengths and
weaknesses.

     d. Social communication. Play is suggested to facili-
tate learning appropriate communicative responses, and developing communication skills (Dolhinow, 1971; Jolly, 1972; Mason,
1965; Poirier and Smith, 1974; Rumbaugh, 1974). Fedigan
(1972) notes that the development of social perception, the
ability to predict another's behavior and respond accordingly,
is a fundamental social skill of primates and may develop in
playful interactions (Poirier and Smith, 1974).

     e. Social integration. Etkin (1967) has suggested that
play may be a method whereby animals maintain social familiarity with other individuals. Play facilitates an individual's
integration into the troop structure (Rosenblum and Lowe,
1971; Southwick, Beg and Siddiqi, 1965), and formation of social bonds (Carpenter, 1934; Jay, 1965; Poirier, 1969, 1970,
1972; Suomi and Harlow, 1971). Furthermore, Poirier (1972)
emphasizes that play functions to enable the individual to
learn the limitations of self-assertiveness, clearly important in proper social integration.

     These functions which have been suggested for play, al-
though appealing and interesting, are based on comparatively
little data (Beach, 1945; Müller-Schwarze, 1971; Welker,
1971). Loizos (1967) has suggested that it is presently more

---

[3]*See Symons (1978) for a critical discussion of this proposition.*

important to conduct precise, systematic observations of play
than to propose additional hypotheses. More data are needed
to evaluate the adaptiveness of play, and, as Baldwin and Baldwin (1977) note, there are alternative avenues to normal social development. Furthermore, Welker (1971) points out that
play should not be considered totally adaptive on a *priori*
grounds, solely because it is a common activity.

It is important to realize that play can have disadvanta-
ges and maladaptive consequences for the individual. Thorpe
(1956:86) postulated that, "Provided, then, that the conditions of life are easy, play, however great in its practical
value may be a means of learning about and so mastering the
external world, is always in danger of becoming the main out-
let for the animal's energies, and so dysgenic."

Baldwin and Baldwin (1977) suggest that play may be maladaptive in some respects, for individuals may be placed in
dangerous or risky situations. Berger (1972) found that juvenile male olive baboons suffered the highest mortality as a
result of their exploratory play behavior and of being driven
from the troop by dominant males. Additionally, young primates are, in many cases, exposed to higher risks than other
age classes (Poirier, 1972).

It has been noted (Baldwin and Baldwin, 1977) that maladaptive learning experiences can result from playful interactions. This has been effectively demonstrated in a wide
variety of species, including humans (Byrd, 1972; McKearney,
1969; Weiner, 1965, 1969), particularly in laboratory situations, although not in the field situation. Statements about
the adaptiveness of particular behaviors (i.e., play) are
mostly speculative since there is very little direct data by
which to evaluate them (Rowell, 1972). Welker (1971:189)
notes, "...none of these views regarding the adaptive function
of play constitutes an explanation in the scientific sense;
rather, they appear as very general hypotheses that have not
been verified or are not testable." Fagen (1974:852) concludes, "Evidence that play facilitates generic learning, or
indeed any demonstrable function, is difficult to obtain."

Frequently, researchers rely on play deprivation studies
(Chepko, 1971; Harlow, Harlow and Hansen, 1963; Müller-
Schwarze, 1968; Oakley and Reynolds, 1976) to demonstrate the
functional significance of play; however, their results are
confounded by behavioral and/or social deprivation of other
types (Bekoff, 1976b); e.g., motor activity (Chepko, 1971),
peer contact (Harlow, Harlow and Hansen, 1963; Marler and
Hamilton, 1966). Dolhinow and Bishop (1971:15) note, "The
problem remains whether it is peer contact or the act of playing that results in normal behavior, and this would be very
difficult to test experimentally."

Deprivation studies are further confounded by data on

squirrel monkeys in Western Panama (Baldwin and Baldwin, 1973, 1974). During a 10-week study of *Saimiri* at Barqueta, no playful behavior was recorded; however, these animals were reported to have exhibited grossly normal behavior patterns. In spite of the lack of play, animals maintained close individual distances and functioned as a cohesive troop, although Poirier (1969) noted that Nilgiri langur troops *(Presbytis johnii)*, which exhibited little social play, fissioned. Baldwin and Baldwin (1973, 1974, 1976) suggest the absence of play was due to a dearth of foods preferred by squirrel monkeys *(Saimiri sciureus)*, a failure to satisfy all primary needs (Meyer-Holzapfel, 1956). Furthermore, Baldwin and Baldwin (1973) report these animals spent 95% of each 14-hour day in foraging and traveling. Clearly, there is a growing body of literature which strongly suggests normal social development is possible in the absence of play; however, Baldwin and Baldwin (1973: 379-380) conclude that, "...the opportunity to play provides learning experiences in which young animals can develop more complex, varied social interaction patterns and stronger habits for engaging in frequent overt exchanges."

The main problem in the functional approach to the study of play is how to test any of the suggested functions of play. In other words, how can one develop a null hypothesis and test it? Fagen (1976) has recently reviewed the "exercise" or physical training hypothesis and has made several physiologically based predictions that should prove useful to researchers in the study of play. Alternatively, the social development, social integration and related social functions might be tested by raising young animals in a social group with procedures that would prevent play [e.g., a "slow feeder" apparatus (Baldwin and Baldwin, 1976)] without depriving the animals of social contact, sensory stimulation, exercise, etc. Without these kinds of empirical tests, the functions of play will remain, at best, educated guesses.

IV. CONCLUSION

In sum, there are different views of play which must be considered if a real understanding of play is to emerge. Considerable care must be exercised so as not to confuse these theoretical positions, as this can only lead to further problems. Clearly, it is out of a historical milieu that has been guilty of anthropomorphism that new insights into play must be acquired. Hansen (1974:183) noted, "...that investigations done on play activities in primates seem to have barely scratched the surface with respect to the potential knowledge

that is available concerning the particulars of how play activity may involve the key to understanding behavioral development."

It should be clear that regardless of theoretical position
(structuralist vs. functionalist), or alternative, the only
way in which the study of play can make any real advances is
through the development of well planned experiments which explicitly state and test various hypotheses. Until this is
done, those who study play will be guilty of proof by assertion, and not empirical verification.

REFERENCES

Aldis, O. "Play Fighting". Academic Press, New York (1975).
Alexander, F. A contribution to the theory of play. *Psychiat.
    Q. 27*, 175-193 (1958).
Altman, J. "Organic Foundations of Animal Behavior". Holt,
    Rinehart and Winston, New York (1966).
Altmann, S. A. A field study of the sociobiology of rhesus
    monkeys, *Macaca mulatta. Ann. N. Y. Acad. Sci. 102*,
    338-435 (1962).
_____ The structure of primate social communication, *in* "So-
    cial Communication Among Primates" (S. A. Altmann, ed.),
    pp. 325-362. University of Chicago Press, Chicago (1967).
Andrew, R. J. Arousal and the causation of behaviour. *Behav-
    iour 51*, 135-165 (1974).
Appleton, L. E. "A Comparative Study of the Play of Adult
    Savages and Civilized Children". University of Chicago
    Press, Chicago (1910).
Baldwin, J. D. The ontogeny of social behavior of squirrel
    monkeys *(Saimiri sciureus)* in a semi-natural environment.
    Folia primat. 11, 35-79 (1969).
Baldwin, J. D., and Baldwin, J. I. The ecology and behavior
    of squirrel monkeys *(Saimiri oerstedi)* in a natural forest
    in Western Panama. *Folia primat. 18*, 161-184 (1972).
_____ The role of play in social organization: comparative
    observations on squirrel monkeys *(Saimiri)*. *Primates 14*,
    369-381 (1973).
_____ Exploration and social play in squirrel monkeys *(Sai-
    miri)*. *Amer. Zool. 14*, 303-314 (1974).
_____ Effects of food ecology on social play: a laboratory
    simulation. *Z. Tierpsychol. 40*, 1-14 (1976).
_____ The role of learning phenomena in the ontogeny of ex-
    ploration and play, *in* "Biosocial Development in Primates:
    A Handbook" (S. Chevalier-Skolnikoff, and F. E. Poirier,
    eds.), pp. 343-406. Garland Publishing Co., New York
    (1977).

Bateson, G. The message "this is play", *in* "Group Processes" (B. Schaffner, ed.), pp. 145-242. Macy Foundation, New York (1955a).

_____ A theory of play and fantasy. *Psychiat. Res. Rep. 2,* 39-50 (1955b).

Beach, F. A. Current concepts of play in mammals. *Amer. Nat. 79,* 523-541 (1945).

Bekoff, M. The development of social interaction, play, and metacommunication in mammals: an ethological perspective. *Quart. Rev. Biol. 47,* 412-434 (1972).

_____ Some notes on the history of play behavior. Paper presented at the Fifth Annual Meeting, International Society for the History of Behavioral and Social Sciences, Plattsburg, New York (1973a).

_____ Mammalian social play: an alliterative approach. Paper presented at the 72nd Annual Meeting, American Anthropological Association, New Orleans (1973b).

_____ Social play and play-soliciting by infant canids. *Amer. Zool. 14,* 323-340 (1974).

_____ The communication of play intention: are play signals functional? *Semiotica 15,* 231-240 (1975).

_____ Animal play: problems and perspectives, *in* "Perspectives in Ethology, Vol. 2" (P. P. G. Bateson, and P. H. Klopfer, eds.), pp. 164-188. Plenum Publishing Co., New York (1976a).

_____ The social deprivation paradigm: who's being deprived of what? *Develop. Psychobiol. 9,* 499-500 (1976b).

Berger, M. E. Population structure of olive baboons *(Papio anubis)* in the Laikipia district of Kenya. *E. Afr. Wildl. J. 10,* 159-164 (1972).

Berlyne, D. E. "Conflict, Arousal, and Curiosity". McGraw-Hill, New York (1960).

_____ Laughter, humor, and play, *in* "Handbook of Social Psychology, Vol. 3" (G. Lindzey, and E. Aronson, eds.), pp. 795-852. Addison-Wesley, Reading, Pennsylvania (1969).

Bertrand, M. The behavioral repertoire of the stumptail macaque. A descriptive and comparative study. *Bibl. primat. 11,* 1-123 (1969).

Bierens de Haan, J. A. Das Spiel eines jungen solitären Schimpansen. *Behaviour 4,* 144-156 (1952).

Bindra, D. "Motivation: A Systematic Reinterpretation" Ronald Press Co., New York (1959).

Bingham, H. C. Parental play of chimpanzees. *J. Mammal. 8,* 77-89 (1927).

Birch, H. G. The relation of previous experience to insightful problem-solving. *J. comp. Physiol. Psychol. 38,* 367-383 (1945).

Breuggeman, J. A. "Adult Play Behavior and Its Occurrence Among Free-Ranging Rhesus Monkeys *(Macaca mulatta)*".

Ph.D. Dissertation, Northwestern University, Evanston (1976).

Brownlee, A. Play in domestic cattle: an analysis of its nature. *Br. Vet. J. 110,* 46-68 (1954).

Bruner, J. S. "Beyond the Information Given: Studies in the Psychology of Knowing". Norton, New York (1973a).

_____ Organization of early skilled action. *Child Develop. 44,* 1-11 (1973b).

Buhler, K. "The Mental Development of the Child". Harcourt Brace, New York (1930).

Buytendijk, F. J. J. "Wesen und Sinn des Spieles". K. Wolff, Berlin (1934).

Byrd, L. D. Responding in the squirrel monkey under second-order schedules of shock delivery. *J. exp. Anal. Behav. 18,* 155-167 (1972).

Carpenter, C. R. A field study of the behavior and social relations of howling monkeys. *Comp. Psychol. Monogr. 10,* 1-168 (1934).

Carr, H. A. The survival value of play, *in* "Investigations of the Department of Psychology and Education". University of Colorado, Boulder (1902).

Chepko, B. D. A preliminary study of the effects of play deprivation on young goats. *Z. Tierpsychol. 28,* 517-526 (1971).

Chevalier-Skolnikoff, S. Facial expression of emotion in non-human primates, *in* "Darwin and Facial Expression: A Century of Research in Review" (P. Ekman, ed.), pp. 11-89. Academic Press, New York (1973).

_____ The primate play face: a possible key to the determinants and evolution of play. *Rice Univ. Stud. 60,* 9-29 (1974).

Cooper, J. B. An exploratory study on African lions. *Comp. Psychol. Monogr. 17,* 1-48 (1942).

Csikszentmihalyi, M., and Bennett, S. An exploratory model of play. *Amer. Anthrop. 73,* 45-58 (1971).

Curti, M. W. "Child Psychology". Longmans, Green and Co., New York (1930).

Darling, F. F. "A Herd of Red Deer". Oxford University Press, London (1937).

Dobzhansky, T. "Mankind Evolving". Yale University Press, New Haven (1962).

Döhl, J., and Podolczak, D. Versuche zur Manipulierfreudigkeit von zwei jungen Orang-Utans *(Pongo pygmaeus)* im Frankfurter Zoo. *Zool. Gart., Frankf., 43,* 81-94 (1973).

Dolan, K. J. Metacommunication in the play of a captive group of Syke's monkeys. Paper presented at the 45th Annual Meeting, American Association of Physical Anthropologists, St. Louis (1976).

Dolhinow, P. J. At play in the fields. *Nat. Hist., N.Y., 80,*
    66-71 (1971).
Dolhinow, P. J., and Bishop, N. The development of motor
    skills and social relationships among primates through
    play. *Minnesota Symp. Child Psychol. 4,* 141-198 (1970).
Drickamer, L. C., and Vessey, S. H. Group changing in free-
    ranging male rhesus monkeys. *Primates 14,* 359-368 (1973).
Eibl-Eibesfeldt, I. Concepts of ethology and their signifi-
    cance in the study of human behavior, *in* "Early Behavior"
    (H. W. Stevenson, E. H. Hess, and H. L. Rheingold, eds.),
    pp. 127-146. John Wiley and Sons, New York (1967).
_____ "Ethology, the Biology of Behavior". Holt, Rinehart
    and Winston, New York (1975).
Erikson, E. H. "Childhood and Society". Norton, New York
    (1950).
_____ Sex differences in the play configurations of preado-
    lescents. *Am. J. Orthopsychiat. 21,* 667-692 (1951).
Etkin, W. "Social Behavior from Fish to Man". University of
    Chicago Press, Chicago (1967).
Ewer, R. F. "Ethology of Mammals". Plenum Press, New York
    (1968).
Fagen, R. M. Selective and evolutionary aspects of animal
    play. *Amer. Nat. 108,* 850-858 (1974).
_____ Exercise, play and physical training in animals, *in*
    "Perspectives in Ethology, Vol. 2" (P. P. G. Bateson, and
    P. H. Klopfer, eds.), pp. 189-219. Plenum Press, New York
    (1976).
Farentinos, R. C. Some observations on the play behavior of
    the stellar sea lion *(Eumetopias jubata).* *Z. Tierpsychol.*
    28, 428-438 (1971).
Fedigan, L. Social and solitary play in a colony of vervet
    monkeys *(Cercopithecus aethiops).* *Primates 13,* 347-364
    (1972).
Fox, M. W. Ontogeny of prey-killing behavior in *Canidae.*
    Behaviour 35, 259-272 (1969).
_____ "Behaviour of Wolves, Dogs and Related Canids". Harper
    and Row, New York (1971).
Freeman, H. E., and Alcock, J. Play behaviour of a mixed group
    of juvenile gorillas and orang-utans *(Gorilla g. gorilla*
    and *Pongo p. pygmaeus).* *Int. Zoo Yb. 13,* 189-194 (1973).
Freud, S. Beyond the pleasure principal, *in* "The Standard
    Edition of the Complete Psychological Works of Sigmund
    Freud, Vol. XVIII" (J. Strachey, ed.), pp. 7-64. Hogarth,
    London (1955).
_____ Creative writers and day dreaming, *in* "The Standard
    Edition of the Complete Psychological Works of Sigmund
    Freud, Vol. XIX" (J. Strachey, ed.), pp. 141-154. Hogarth,
    London (1959a).
_____ Inhibitions, symptoms and anxiety, *in* "The Standard

Edition of the Complete Psychological Works of Sigmund
Freud, Vol. XX" (J. Starchey, ed.), pp. 177-178. Hogarth,
London (1959b).

Gehlen, A. "Der Mensch seine Natur und Seine Stellung in der
Welt". Athenäum Verlag, Frankfurt (1940).

Geist, V. "Mountain Sheep: A Study of Behavior and Evolution". University of Chicago Press, Chicago (1971).

Gentry, R. L. The development of social behavior through play
in the stellar sea lion. *Amer. Zool. 14,* 391-403 (1974).

Ghosh, S. Play instinct. *Indian J. Psychol. 10,* 159-162
(1935).

Gilmore, J. B. "The Role of Anxiety and Cognitive Factors in
Children's Play Behavior". Ph.D. Dissertation, Yale University, New Haven (1964).

_____ Play: a special behavior, *in* "Current Research in
Motivation" (R. N. Haber, ed.), pp. 343-354. Holt, Rinehart and Winston, New York (1966).

Goodall, J. Chimpanzees of the Gombe Stream Reserve, *in* "Pri-
mate Behavior: Field Studies of Monkeys and Apes" (I.
DeVore, ed.), pp. 425-473. Holt, Rinehart and Winston,
New York (1965).

Gordon, T. P., Rose, R. M., and Bernstein, I. S. Seasonal
rhythm in plasma testosterone levels in the rhesus monkey
(Macaca mulatta): a three year study. *Horm. Behav. 7,*
229-243 (1976).

Groos, K. "Play of Animals". Appleton, New York (1898).
_____ "The Play of Man". Appleton, New York (1908).

Gulick, L. J. "A Philosophy of Play". Charles Scribner's
Sons, New York (1920).

Gwinner, E. Über einige Beswegungsspiele des Kolkraben. *Z.
Tierpsychol. 23,* 28-36 (1966).

Hall, G. S. "Youth" Its Education, Regimen and Hygiene".
Appleton, New York (1906).

Hall, K. R. L. Variations on the ecology of the chacma baboon
(P. ursinus). Symp. zool. Soc. Lond. 10, 1-28 (1963).

Hall, K. R. L., and DeVore, I. Baboon social behavior, *in*
"Primate Behavior: Field Studies of Monkeys and Apes"
(I. DeVore, ed.), pp. 53-110. Holt, Rinehart and Winston,
New York (1965).

Hansen, E. W. The development of maternal and infant behavior
in the rhesus monkey. *Behaviour 27,* 107-149 (1966).

_____ Some aspects of behavioural development in evolutionary
perspective, *in* "Ethology and Psychiatry" (N. F. White,
ed.), pp. 182-186. University of Toronto Press, Toronto
(1974).

Harlow, H. F., and Harlow, M. K. The affectional systems, *in*
"Behavior of Nonhuman Primates, Vol. 2" (A. M. Schrier,
H. F. Harlow, and F. Stollnitz, eds.), pp. 287-334. Aca-
demic Press, New York (1965).

_____ Effects of various mother-infant relationships on rhesus monkey behaviors, *in* "Determinants of Infant Behavior, Vol. 4" (B. M. Foss, ed.), pp. 15-36. Methuen, London (1969).

Harlow, H. F., Harlow, M. K., and Hansen, E. W. The maternal effective system of rhesus monkeys, *in* "Maternal Behavior in Mammals" (H. L. Rheingold, ed.), pp. 254-281. John Wiley and Sons, New York (1963).

Hebb, D. O. "The Organization of Behavior". John Wiley and Sons, New York (1949).

_____ Drives and the CNS (conceptual nervous system). *Psychol. Rev. 62,* 243-254 (1955).

Heckhausen, H. Entwurf einer Psychologie des Spielens. *Psychol. Forsch. 27,* 225-243 (1964).

Henry, J. D., and Herrero, S. M. Social play in the American black bear: its similarity to canid social play and an examination of its identifying characteristics. *Amer. Zool. 14,* 371-389 (1974).

Herron, R. E., and Sutton-Smith, B. (eds.) "Child's Play". John Wiley and Sons, New York (1971).

Hinde, R. A. Development of social behavior, *in* "Behavior of Nonhuman Primates, Vol. 3" (A. M. Schrier, and F. Stollnitz, eds.), pp. 1-68. Academic Press, New York (1971).

Hinton, H. E., and Dunn, A. M. S. "Mongooses: Their Natural History Behaviour". Oliver and Boyd, London (1967).

Hurlock, E. B. Experimental investigations of childhood play. *Psychol. Bull. 31,* 47-66 (1934).

Hutt, C. Exploration and play in children. *Symp. zool. Soc. Lond. 18,* 61-81 (1966).

_____ Specific and diversive exploration. *Adv. Child Dev. Behav. 5,* 119-180 (1970).

Jay, P. C. Mother-infant relations in langurs, *in* "Maternal Behavior in Mammals" (H. L. Rheingold, ed.), pp. 282-304. John Wiley and Sons, New York (1963).

_____ The common langur of north India, *in* "Primate Behavior: Field Studies of Monkeys and Apes" (I. DeVore, ed.), pp. 197-249. Holt, Rinehart and Winston, New York (1965).

Jolly, A. "Lemur Behavior". University of Chicago Press, Chicago (1966).

_____ "The Evolution of Primate Behavior". MacMillan Co., New York (1972).

Kaufmann, J. H. Social relations of adult males in a free-ranging band of rhesus monkeys, *in* "Social Communication Among Primates" (S. A. Altmann, ed.), pp. 73-98. University of Chicago Press, Chicago (1967).

Koehler, O. Vom Spiel bei Tieren. *Freiburger Dies Universitatis 13,* 1-32 (1966).

Koford, C. B. Ranks of mothers and sons in bands of rhesus monkeys. *Science 141,* 356-357 (1965).

Kruuk, H. "The Spotted Hyena: A Study of Predation and Social Behavior". University of Chicago Press, Chicago
(1972).

Kummer, H. "Primate Societies: Group Techniques of Ecological Adaptation". Aldine, Chicago (1971).

Lancaster, J. B. Play-mothering: the relations between juvenile females and young infants among free-ranging vervet
monkeys *(Cercopithecus aethiops)*. *Folia primat. 15,*
337-363 (1971).

Lazar, J. W., and Beckhorn, G. D. The concept of play and the
development of social behavior in ferrets *(Mustela putori-
ous)*. *Amer. Zool. 14,* 405-414 (1974).

Lazarus, M. "Uber die Reize des Spiels". F. Dummler, Berlin
(1883).

Lehman, H. C., and Witty, P. A. "The Psychology of Play Activities". A. S. Barnes and Co., New York (1927).

Leuba, C. Toward some integration of learning theories: the
concept of optimal stimulation. *Psychol. Rep. 1,* 27-33
(1955).

Levitsky, D. A., and Barnes, R. H. Nutritional and environmental interactions in the behavioral development of the
rat: long term effects. *Science 176,* 68-71 (1972).

Lewin, K. Environmental forces, *in* "A Handbook of Child Psychology" (C. Murchinson, ed.), pp. 590-625. Clark University Press, Worcester, Massachusetts (1933).

Lichstein, L. "Play in Rhesus Monkeys: I) Definition. II)
Diagnostic Significance". Ph.D. Dissertation, University
of Wisconsin, Madison (1973a).

_____ Play in rhesus monkeys: I) definition. II) diagnostic
significance. *Diss. Abstr. Intl. 33,* 3985 (1973b).

Loizos, C. Play in mammals. *Symp. zool. Soc. Lond. 18,* 1-9
(1966).

_____ Play behavior in higher primates: a review, *in* "Primate Ethology" (D. Morris, ed.), pp. 176-218. Aldine,
Chicago (1967).

Lorenz, K. Z. The comparative method in studying innate behaviour patterns. *Symp. Soc. exp. Biol. 4,* 221-268 (1950).

_____ Plays and vaccum activities in animals, *in* "L'instinct
dans le Compartement des Animaux et de l'Homme" (P. P.
Grasse, ed.), pp. 633-645. Fondation Singer-Polignac,
Paris (1956).

Loy, J. Behavioral responses in free-ranging rhesus monkeys
to food shortages. *Am. J. phys. Anthrop. 33,* 263-272
(1970).

Loy, J., and Loy, K. Behavior of an all-juvenile group of
rhesus monkeys. *Am. J. phys. Anthrop. 40,* 83-96 (1974).

Marler, P., and Hamilton, W. J. "Mechanisms of Animal Behavior". John Wiley and Sons, New York (1966).

Marsden, H. M. Agonistic behaviour of young rhesus monkeys

after changes induced in social rank of their mothers. *Anim. Behav. 16*, 38-44 (1968).

Mason, W. A. The social development of monkeys and apes, *in* "Primate Behavior: Field Studies of Monkeys and Apes" (I. DeVore, ed.), pp. 514-543. Holt, Rinehart and Winston, New York (1965).

_____ Motivational aspects of social responsiveness in young chimpanzees, *in* "Early Behavior - Comparative and Developmental Approaches" (H. W. Stevenson, E. H. Hess, and H. L. Rheingold, eds.), pp. 103-126. John Wiley and Sons, New York (1967).

_____ Early social deprivation in the nonhuman primates: implications for human behavior, *in* "Environmental Influences" (D. Glass, ed.), pp. 70-101. Russell Sage Foundation, New York (1968).

_____ Motivational factors in psychosocial development, *in* "Nebraska Symposium on Motivation" (W. J. Arnold, and M. M. Page, eds.), pp. 35-67. University of Nebraska Press, Lincoln (1971).

McKearney, J. W. Fixed-interval schedules of electric shock presentation: extinction and recovery of performance under different shock intensities and fixed-interval durations. *J. exp. Anal. Behav. 12*, 301-313 (1969).

Mears, C. E., and Harlow, H. F. Play: early and eternal. *Proc. nat. Acad. Sci.,U.S.A. 72*, 1878-1882 (1975).

Meier, G. W., and Devanney, V. D. The ontogeny of play within a society: preliminary analysis. *Amer. Zool. 14*, 289-294 (1974).

Meyer-Holzapfel, M. Das Spiel bei Säugetieren. *Handb. Zool. 8*, 1-36 (1956).

Millar, S. "The Psychology of Play". Pelican, London (1968).

Miller, S. Ends, means, and galumphing: some leitmotifs of play. *Amer. Anthrop. 75*, 87-98 (1973).

Mitchell, C. P. "The Childhood of Animals". Frederick A. Stokes, New York (1912).

Mitchell, E. D., and Mason, B. S. "The Theory of Play". A. S. Barnes, New York (1934).

Mitchell, G., and Brandt, E. M. Paternal behavior in primates, *in* "Primate Socialization" (F. E. Poirier, ed.), pp. 173-206. Random House, New York (1972).

Morgan, C. L. "Animal Behavior". Edward Arnold, London (1900).

Morris, D. "The Biology of Art". Metheun, London (1962).

Müller-Schwarze, D. Play deprivation in deer. *Behaviour 31*, 144-162 (1968).

_____ Ludic behavior in young mammals, *in* "Brain Development and Behavior" (M. B. Sterman, D. J. McGinty, and A. M. Adinolfi, eds.), pp. 229-249. Academic Press, New York (1971).

Oakley, F. B., and Reynolds, P. C. Differing responses to
     social play deprivation in two species of macaque, *in* "The
     Anthropological Study of Play: Problems and Perspectives"
     (D. F. Lancy, and B. A. Tindall, eds.), pp. 179-188.
     Leisure Press, Cornwall, New York (1976).
Owens, N. W. Social play behaviour in free-living baboons,
     Papio anubis. Anim. Behav. 23, 387-408 (1975).
Patrick, G. T. W. "The Psychology of Relaxation". Houghton
     Mifflin, New York (1916).
Piaget, J. "Play, Dreams, and Imitation in Childhood". Wil-
     liam Heinemann, Ltd., London (1951).
Poirier, F. E. The Nilgiri langur troop: its composition,
     structure, function, and change. *Folia primat. 19,* 20-47
     (1969).
_____ The Nilgiri langur of north India, *in* "Primate Behav-
     ior: Developments in Field and Laboratory Research,
     Vol. 1" (L. A. Rosenblum, ed.), pp. 254-283. Academic
     Press, New York (1970).
_____ Introduction, *in* "Primate Socialization" (F. E. Poir-
     ier, ed.), pp. 3-28. Random House, New York (1972).
Poirier, F. E., and Şmith, E. O. Socializing functions of
     primate play. *Amer. Zool. 14,* 275-287 (1974).
Poole, T. B. Aggressive play in polecats. *Symp. zool. Soc.
     Lond. 18,* 23-44 (1966).
Poole, T. B., and Fish, J. An investigation of playful behav-
     iour in *Rattus norvegicus* and *Mus musculus* (Mammalia).
     J. zool. Soc. Lond. 175, 61-71 (1975).
Pycraft, W. F. "The Infancy of Animals". Henry Holt and Co.,
     New York (1913).
Reaney, M. J. The psychology of the organized group game.
     Brit. J. Psychol., Monogr. Suppl. 1, 1-76 (1916).
Redican, W. K., and Mitchell, G. Play between adult male and
     infant rhesus monkeys. *Amer. Zool. 14,* 295-302 (1974).
Rensch, B. Play and art in apes and monkeys, *in* "Symposia of
     the Fourth International Congress of Primatology, Vol. 1:
     Precultural Primate Behavior" (E. W. Menzel, ed.), pp.
     102-123. Karger, Basel (1973).
Rensch, B., and Dücker, G. Spiele von Mungo und Ichneumon.
     Behaviour 14, 185-213 (1959).
Reynolds, V. "The Apes". E. P. Dutton and Co., New York
     (1967).
Rheingold, H. L. Introduction, *in* "Maternal Behavior in Mam-
     mals" (H. L. Rheingold, ed.), pp. 1-8. John Wiley and
     Sons, New York (1963).
Rhine, R. J. Variation and consistency in the social behavior
     of two groups of stumptail macaques *(Macaca arctoides).
     Primates 14,* 21-36 (1973).
Riesen, A. H. Stimulation as a requirement for growth and
     function in behavioral development, *in* "Functions of

Varied Experience" (D. W. Fiske, and S. R. Maddi, eds.), pp. 57-80. Dorsey Press, Homewood, Illinois (1961).

_____ Effects of early deprivation of photic stimulation, *in* "The Biosocial Basis of Mental Retardation" (S. F. Osler, and R. E. Cooke, eds.), pp. 61-85. Johns Hopkins Press, Baltimore (1965).

Robinson, E. S. The compensatory function of make believe play. *Psychol. Rev. 27,* 429-439 (1920).

Rosenblum, L. A., Kaufman, I. C., and Stynes, A. J. Interspecific variations in the effects of hunger on diurnally varying behavior elements in macaques. *Brain, Behav. Evol. 2,* 119-131 (1969).

Rosenblum, L. A., and Lowe, A. The influence of familiarity during rearing on subsequent partner preferences in squirrel monkeys. *Psychon. Sci. 23,* 35-37 (1971).

Rowell, T. E. "The Social Behavior of Monkeys". Penguin, Baltimore (1972).

Rumbaugh, D. M. Comparative primate learning and its contribution to understanding development, play, intelligence, and language. *Advanc. behav. Biol. 9,* 253-281 (1974).

Sade, D. S. Determinants of dominance in a group of freeranging rhesus monkeys, *in* "Social Communication Among Primates" (S. A. Altmann, ed.), pp. 99-114. University of Chicago Press, Chicago (1967).

_____ An ethogram for rhesus monkeys. I. Antithetical contrasts in posture and movements. *Am. J. phys. Anthrop. 38,* 537-542 (1973).

Schaller, G. B. "The Mountain Gorilla: Ecology and Behavior". University of Chicago Press, Chicago (1963).

Schenckel, R. Play, exploration, and territoriality in the wild lion. *Symp. zool. Soc. Lond. 18,* 11-22 (1966).

Schiller, F. "Essays, Aesthetical and Philosophical". Bell and Sons, London (1875).

Schiller, P. H. Innate motor action as a basis for learning: manipulative patterns in the chimpanzee, *in* "Instinctive Behavior" (C. Schiller, ed.), pp. 264-287. International Universities Press, New York (1957).

Schlosberg, H. The concept of play. *Psychol. Rev. 54,* 229-231 (1947).

Schneirla, T. C. An evolutionary and developmental theory of biphasic processes underlying approach and withdrawal, *in* "Nebraska Symposium on Motivation" (M. R. Jones, ed.), pp. 1-42. University of Nebraska Press, Lincoln (1959).

Simonds, P. E. The bonnet macaque of South India, *in* "Primate Behavior: Field Studies of Monkeys and Apes" (I. DeVore, ed.), pp. 175-196. Holt, Rinehart and Winston, New York (1965).

Smith, E. O. "Social Play in Rhesus Macaques *(Macaca mulatta)*". Ph.D. Dissertation, Ohio State University, Columbus (1977).

Southwick, C. H.   An experimental study of intragroup agonis-
    tic behavior in rhesus monkeys *(Macaca mulatta)*. *Behav-
    iour 28*, 182-209 (1967).
Southwick, C. H., Beg, M. A., and Siddiqi, M. R.   Rhesus mon-
    keys in north India, *in* "Primate Behavior: Field Studies
    of Monkeys and Apes" (I. DeVore, ed.), pp. 111-159. Holt,
    Rinehart and Winston, New York (1965).
Spencer, H.   "Principles of Psychology". Appleton, New York
    (1873).
Steiner, A. L.   Play activity of Columbian ground squirrels.
    Z. Tierpsychol. 28, 247-261 (1971).
Struhsaker, T. T.   Auditory communication among vervet monkeys
    (Cercopithecus aethiops), *in* "Social Communication Among
    Primates" (S. A. Altmann, ed.), pp. 281-324. University
    of Chicago Press, Chicago (1967).
Suomi, S. J., and Harlow, H. F.   Monkeys at play. *Nat. Hist.*,
    New York, 80, 72-75 (1971).
Sutton-Smith, B.   Conclusion, *in* "Child's Play" (R. E. Herron,
    and B. Sutton-Smith, eds.), pp. 343-345. John Wiley and
    Sons, New York (1971).
Symons, D.   "Aggressive Play in a Free-Ranging Group of Rhesus
    Macaques *(Macaca mulatta)*". Ph.D. Dissertation, Univer-
    sity of California, Berkeley (1973).
_____   "Play and Aggression: A Study of Rhesus Monkeys".
    Columbia University Press, New York (1978).
Tembrock, G.   Spielverhalten und vergleichende Ethologie. Beo-
    bachtungen zum Spiel von *Alopex lagopus*. *Z. Säugetierk.*
    25, 1-14 (1960).
Thorpe, W. H.   "Learning and Instinct in Animals". Metheun,
    London (1956).
Tinklepaugh, O. L.   Social psychology of animals, *in* "Compara-
    tive Psychology" (F. A. Moss, ed.), pp. 449-482. Pren-
    tice-Hall, New York (1934).
Tolman, E. C.   "Purposive Behavior in Animals and Man". Cen-
    tury Co., New York (1932).
Tsumori, A.   Newly acquired behavior and social interactions
    of Japanese monkeys, *in* "Social Communication Among Pri-
    mates" (S. A. Altmann, ed.), pp. 207-219. University of
    Chicago Press, Chicago (1967).
van Hooff, J. A. R. A. M.   The facial displays of catarrhine
    monkeys and apes, *in* "Primate Ethology" (D. Morris, ed.),
    pp. 7-68. Aldine, Chicago (1967).
_____   A comparative approach to the phylogeny of laughter
    and smiling, *in* "Non-verbal Communication" (R. A. Hinde,
    ed.), pp. 209-241. Cambridge University Press, Cambridge
    (1972).
van Lawick-Goodall, J.   Mother-offspring relationships in
    free-ranging chimpanzees, *in* "Primate Ethology" (D. Mor-
    ris, ed.), pp. 287-346. Aldine, Chicago (1967).

_____ The behavior of free-living chimpanzees in the Gombe Stream Reserve. *Anim. Behav. Monogr. 1,* 165-311 (1968).

Vessey, S. H. Free-ranging rhesus monkeys: behavioral effects of removal, separation and reintroduction of group members. *Behaviour 40,* 216-227 (1971).

Volkman, F. R., and Greenough, W. T. Rearing complexity affects branching of dendrites in the visual cortex of the rat. *Science 176,* 1445-1447 (1972).

Washburn, S. L. Primate field studies and social science, *in* "Cultural Illness and Health: Essays in Human Adaptation" (L. Nader, and T. W. Maretzki, eds.), pp. 128-134. American Anthropological Association, Washington (1973).

Washburn, S. L., and Hamburg, D. The implications of primate research, *in* "Primate Behavior: Field Studies of Monkeys and Apes" (I. DeVore, ed.), pp. 607-622. Holt, Rinehart and Winston, New York (1965).

Weiner, H. Conditioning history and maladaptive human operant behavior. *Psychol. Rep. 17,* 935-942 (1965).

_____ Controlling human fixed interval performance. *J. exp. Anal. Behav. 12,* 349-373 (1969).

Welker, W. I. Some determinants of play and exploration in chimpanzees. *J. comp. Physiol. Psychol. 49,* 84-89 (1954).

_____ Effects of age and experience on play and exploration of young chimpanzees. *J. comp. Physiol. Psychol. 49,* 223-226 (1956a).

_____ Variability of play and exploratory behavior in chimpanzees. *J. comp. Physiol. Psychol. 49,* 181-185 (1956b).

_____ Genesis of exploratory and play behavior in infant raccoons. *Psychol. Rep. 5,* 764 (1959).

_____ An analysis of exploratory and play behavior in animals, *in* "Functions of Varied Experience" (D. W. Fiske, and S. R. Maddi, eds.), pp. 278-325. Dorsey Press, Homewood, Illinois (1961).

_____ Ontogeny of play and exploration: a definition of problems and a search for new conceptual solutions, *in* "The Ontogeny of Vertebrate Behavior" (H. Moltz, ed.), pp. 171-228. Academic Press, New York (1971).

Wemmer, C., and Fleming, M. J. Ontogeny of playful contact in a social mongoose, the meerkat, *Suricata suricatta. Amer. Zool. 14,* 415-426 (1974).

West, M. Social play in the domestic cat. *Amer. Zool. 14,* 427-436 (1974).

White, R. W. Motivation reconsidered: the concept of competence. *Psychol. Rev. 66,* 297-333 (1959).

Wilson, S. The development of social behavior in the vole *(Microtus agrestis). Zool. J. Linn. Soc. 52,* 45-62 (1973).

_____ Juvenile play of the common seal *Phoca vitulina vitulina* with comparative notes on the grey seal *Halichoerus grypus. Behaviour 48,* 37-60 (1974).

Wilson, S. C., and Kleiman, D. G. Eliciting play: a compara-
    tive study *(Octodon, Octodontomys, Pediolagus, Phoca,
    Choeropsis, Ailuropoda)*. *Amer. Zool. 14,* 341-370 (1974).
Winch, W. H. Psychology and philosophy of play. I. *Mind 15,*
    32-52 (1906a).
_____ Psychology and philosophy of play. II. *Mind, 15,* 177-
    190 (1906b).
Winter, P. Social communication in the squirrel monkey, *in*
    "The Squirrel Monkey" (L. A. Rosenblum, and R. W. Cooper,
    eds.), pp. 235-253. Academic Press, New York (1968).
Winter, P., Ploog, D., and Latta, J. Vocal repertoire of the
    squirrel monkey *(Saimiri sciureus)*, its analysis and sig-
    nificance. *Exp. Brain Res. 1,* 359-384 (1966).

SEX DIFFERENCES IN THE ACQUISITION OF PLAY
AMONG JUVENILE VERVET MONKEYS[1]

Claud A. Bramblett

Department of Anthropology
The University of Texas
Austin, Texas

I. INTRODUCTION

Loizos (1967) begins her excellent review of play in
higher primates by pointing out that play is a human concept.
Indeed, one of the remarkable things that characterizes the
behavioral literature is the consistency that untrained observers and seasoned "experts" exhibit in recognizing a complex of behaviors called play. Presumably an analogue is
being made with a fundamentally similar behavior in human beings. This ability to recognize play is echoed throughout the
literature, but the number of truly quantitative studies of
play that treat a longitudinal sample is quite small. This is
especially true when one focuses on a single species, in this
report, the vervet monkey *(Cercopithecus aethiops)*.

Struhsaker (1967) describes a vervet vocalization, *purr,*
which he heard when juveniles were engaged in wrestling and
grappling. In his description of inaudible behaviors, he separates play into categories, social and solitary. Social play
included gamboling (hopping or jumping about), wrestling,
grabbing at, grappling, mouthing, embracing, chasing, counter
chasing, heterosexual and homosexual mounting, tug of war, and
hide-and-seek. He observed only a single incidence of soli-

[1]*Financial assistance from the Wenner-Gren Foundation for
Anthropological Research Grant #2635, Biomedical Sciences Support Grant 5SO 5FR 07091-04, and NSF-USDP GU-1598 provided
animal care, observer salaries and part of the data processing.*

tary play, which consisted of manipulation, carrying, and hopping on and off of a feather.

Fedigan (1971) previously analyzed the types of play observed in the same vervet group as is used in the present study. She defined 26 play behaviors and summarized them into seven general categories: directs aggressive play, receives aggressive play, solitary locomotor play, social locomotor play, receives female chastisement, directs approach-withdrawal play, and individual object manipulation. She observed in social play that (1) adults play with immature animals, (2) adult males frequently engage in rough play with older male juveniles, (3) juvenile males and subadult females frequently stimulate play in young infants, and (4) males tend to play within their peer group.

"Play-mothering" has been used as a label to refer to protective or nurturing behaviors when they are performed by presumed nulliparous juvenile females toward infants (Lancaster, 1971). Estrada and Bernal (1977) utilize seven play categories in their study of vervet social play: invitation, wrestling, boxing, chase, displacement, mount, and separation. Their results parallel those of Fedigan. They observe, in addition, that play has temporal structure, and their data are consistent with Mason's (1965) hypothesis that play is inhibited by stressful conditions.

The specific aims of this paper are threefold: (1) to describe changes in frequency of play behaviors as the participants mature from neonates to adults, (2) to construct a mathematical model which describes these age changes in behavior, and (3) to propose hypotheses about mechanisms which structure or influence age-related behavioral changes that are suggested by the description.

II. MATERIALS AND METHODS

A. *The Study Group*

The vervet monkey colony *(Cercopithecus aethiops)* used in this study was formed in 1966 by T. E. Rowell at Makerere University College, Kampala, Uganda, from animals that were trapped nearby. After a stabilization period, she studied their social behavior and reproductive biology (1970, 1971, 1972, 1974a,b). The study group was moved to the Balcones Research Center at The University of Texas, Austin, in November, 1968. Initial studies concerned reproduction (Bramblett, Pejaver and Drickman, 1975). Behavioral observations continued, but have focused on longitudinal changes and stability in the group rather than a characterization of vervet behav-

iors. The results are a series of data sets extending from
1970. Several papers and brief reports have utilized parts of
the data set (Bramblett, 1970, 1971, 1973, 1974, 1976a,b;
Buckley, 1976; Chickering, 1976; Fedigan, 1971, 1973; Hunter,
1973).

The 18 subadult members (nine male and nine female) of the
vervet colony are represented in Figure 1. Each o indicates
an age/sex class for which a score for play is available.
Four wild-born adults, a male and three females, parents of
these subadults, were also present in the colony, but their
play behaviors are not discussed in the analyses.

B. Procedure

Observations are made in standardized test sessions of
1000 seconds duration. The vervet monkeys are observed from a
maximum distance of 10 m by an observer sitting at a table
just outside the midpoint of the outdoor cage (approximately
4 m wide, 17 m long, and 3.5 m high). Food, water, and shade
are always available in the outdoor cage. The door to the

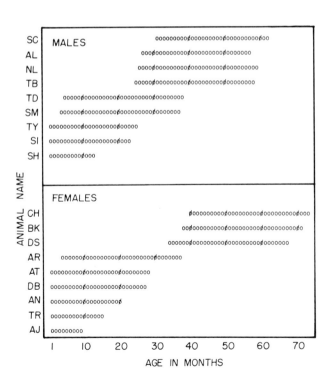

FIGURE 1. Ages of vervet monkeys in this study.

indoor sleeping quarters is closed and locked during testing
to prevent animals from going out of sight. The observer lo-
cates the position of each animal, sets the stopwatch or timer
at 0, and begins the test. Included at the top of each test
page is the date, time, test number, weather, observer's name,
and comments (if unusual events occurred or if the test had to
be terminated).

Play is one of 35 acts on the behavior list. As an act is
perceived (ad libitum) by the observer, it is noted on the recording instrument, a printed page which contains a checklist
of inaudible behavior patterns taken from Struhsaker's (1967)
catalogue. The observer enters abbreviations that identify
actor, recipient (if any), and sequence number. When rapid,
short sequences occur, he waits until a sequence is ended before entering the observations. Long sequences, or major
group disruptions cannot be recorded and result in termination
of the test session. It is not expected, nor is it possible,
for the observer to record every event that occurs, for he is,
himself, a sampling instrument. Reliability of multiple tech-
nicians, each with at least three months previous observation
experience with the group prior to inclusion of his data in
this study, is assessed by comparison of matrices generated
from separate observer's data. Distribution of tests covers
the entire day. The 8,887 play acts analyzed in this sample
are extracted from the 261.9 hours of test data gathered over
a 34-month period.

C. *Analytical Methods*

Behavioral definitions follow Struhsaker (1967). Since
this is a longitudinal description of occurrence of play behaviors from birth to five years, all sub-types of play (except play-mothering, which was not included as a play category
in the checklists used for observation) are combined into a
single act, play. Although actor and recipient of play are
recorded in the original data, in this analysis they are combined. A play session between two animals appears in the
score for each participant. Thus, the score used for this
analysis represents the rate of involvement in social play.

Scores for play behaviors are tabulated and a summary record (containing each subject's name, sex, age, date, uterine
group, number of play event records, and total number of seconds the group was tested during that month) is produced for
each animal for each month. Rates are computed by dividing
the number of play events by the number of seconds of testing.
Since this number is very small for some individuals, all
rates are multiplied by 10^3 to prevent loss of data by computer rounding and are expressed in behaviors per thousand

seconds. Graphs of rate by age provide a visual representation of the tabulated data (Figures 2 and 3). In order to avoid certain computation problems, the time interval between birth and 30 days of age is numbered month one, the second month of life is numbered month two, and so on. There is no zero age.

Mean rates for each month of age are calculated for each sex and graphed in Figure 4. Visually apparent differences in the graphed data are examined with the sign test (Siegel, 1956). The F test, Eq. (1),

$$F = s_1^2 / s_2^2 \tag{1}$$

is used to compare the equality of two variances (Snedecor and Cochran, 1967).

Regression analysis describes a statistical relationship between the age of the individual and rate of play. Methods and procedures are outlined in Ward and Jennings (1973) and discussed at length in Neter and Wasserman (1974). Since the distribution of mean scores for play appears curvilinear, we have chosen a third order polynomial model with one independent variable, age, represented by Eq. (2).

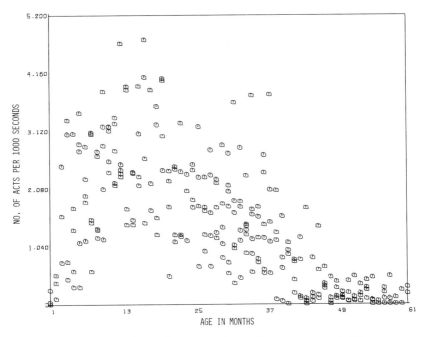

FIGURE 2. Play in male vervet monkeys. o = rate.

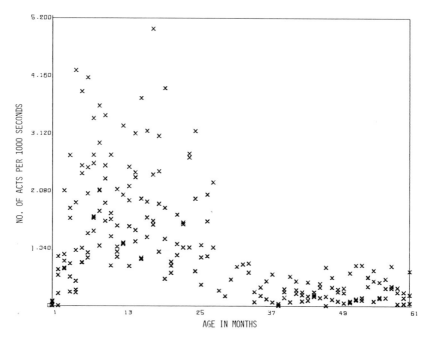

FIGURE 3. Play in female vervet monkeys. x = rate.

$$Y' = a_0 x^0 + a_1 x^1 + a_2 x^2 + a_3 x^3 + e \qquad (2)$$

where Y   =   predicted rate of play

$\quad\quad\quad$ Y' = $\log_{10} Y$

$\quad\quad\quad$ a_j =   regression coefficient

$\quad\quad\quad$ x   =   age in months

$\quad\quad\quad$ e   =   error

The $\log_{10}$ transformation model was chosen and evaluated by
an analysis of the residuals (the difference between observed
and predicted values). Standardized residuals were first
graphed against order of occurrence, predicted values, and
age. These graphs were examined for evidence of a systematic
departure from the regression model. Standardized residuals
were then graphed on a probability plot. Divergence from the
expected patterns in the graphs which suggest divergence of
error variance, inadequate complexity of the model, noninde-
pendence of errors, or non-normality of error would terminate
the use of the $\log_{10}$ model at this stage of analysis. However, if the model passes this graphic analysis, the residuals
are tabulated and subjected to a closer statistical analysis

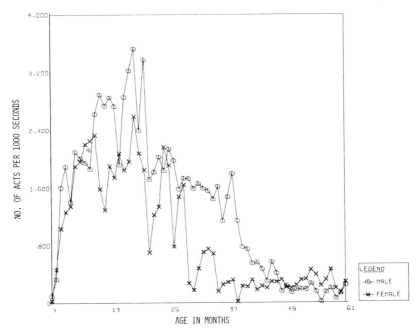

FIGURE 4. Observed mean rates of play in vervet monkeys.
o = male; x = female.

to verify the absence of correlation with the independent variable, randomness, normality, and consistency of variance.   In
this analysis of play, outlying data points are not excluded.
    The polynomial model has no validity for an extended age
range to predict Ys that lie beyond the limits of the observations which form the basis for the model.   In fact, the predictions may become inflated or underestimated as the limits
of observed sample range are approached.   Consequently, all
animals are included in the regression computations which
formulated and tested the model.   The discussion in this chapter is limited, however, to those animals under five years old
(month 1 to month 61).
    Accuracy of the entire regression approximation is as-
sessed by the cumulative F value, the F test for lack of fit,
coefficient of multiple determination, and analysis of residuals.   The F test for lack of fit is based upon separation of
the residual sum of squares into its pure error and its lack
of fit components.   A ratio of lack of fit mean square to pure
error mean square provides an F test for aptness of a regression function.   In this case, a significant F value would
imply an inappropriate model (Neter and Wasserman, 1974).   The
aptness of the model to a particular range of age values is
checked by matching the observed means with predicted values

for that particular range by the sign test. There are direct
computations for comparison of regression lines (Neter and
Wasserman, 1974), but these techniques take into account the
amplitude of the distribution as well as the shape of the line
which connects the predicted means. In order to test the hypothesis that the underlying innate timing mechanisms are
similar in both sexes, we converted computed values to their
proportionate contribution to the total records for that event
by computing the area under the regression line for each month
and dividing that value by the total area for all months.
Thus, a set of computed values independent of the relative activity rate are produced, retaining the shape of the distribution, but placing all behaviors on the same scale. The total
area for all months in the new distribution equals 1.0 for
each sex. The area of the plane bounded by the line x = 1
(age one month) and x = 61 (61 months of age) for the graph of
positive values of the cubic polynomial is defined as the
integral, Eq. (3).

$$\text{AREA} = \int_1^{61} f(s) \, dx \tag{3}$$

The trapezoidal rule gives acceptable approximations of the
integral in Eq. (4).

$$\text{AREA} = \sum_1^{60} \frac{c(f(x_a) + f(x_b))}{} \tag{4}$$

where c = b - a = 1 month

Another, perhaps simpler, algebraic computation is represented
by Eq. (5).

$$\text{AREA} = \sum_1^{60} a_0(x_b - x_a) + \frac{a_1(x_b^2 - x_a^2)}{2} + \frac{a_2(x_b^3 - x_a^3)}{3} + \tag{5}$$

$$\frac{a_3(x_b^4 - x_a^4)}{4}$$

Comparison of values from Eqs. (4) and (5) shows a negligible
difference in the area estimates between the two calculations.
Once the conversion is completed, the new predicted values are
compared for similarity in shape by means of the sign test.

Negative scores were not encountered in the computations
associated with play, but had they occurred, those values
would be rejected and treated as zeroes. Negative predicted
values might make algebraic sense, but behavioral analogues
are difficult to conceptualize.

III. RESULTS

Play is part of the activities of both males and females
from their first month of life. Figures 2 and 3 are graphs of
the observed rates of play for each animal between the ages of
one and 61 months. The mean rate of play in play bouts per
1000 seconds is plotted for both sexes in Figure 4. Between
month 1 and month 47, males play more frequently than females
($Z = 5.25$; $p < .05$). No significant differences in individual
variability by sex were found during this period ($F = 1.04$;
$p > .05$). Conversely, females are involved in play more often
than are males between months 48 and 61 ($Z = 2.94$; $p < .05$).
These older females also exhibit much greater variability in
rate of play than do older males ($F = 1.90$; $p < .05$).

Table I includes the results of the regression computations. A satisfactory F value and coefficient of multiple regression is computed for both sexes. A visual analysis of
residuals does not indicate any noteworthy departures from the
intended regression model. Figure 5 presents the results of

TABLE I. *Regression of Play with Age among Vervet Monkeys*

	Male	Female
A_0	-0.273	-0.0941
A_1	0.0797	0.0282
A_2	-0.00276	-0.00114
A_3	0.0000212	0.00000809
Monkey months	287	337
Degrees of freedom	283	333
Cumulative F	131.98	50.00
P(F)	<0.001	<0.001
R square	0.644	0.738
Adjusted R square	0.641	0.735
Residual SS	29.9955	46.4327
Total SS	84.4806	171.2792
Lack of fit SS	6.0534	1.2408
Lack of fit F	0.63	0.01
P(LFF)	>0.05	>0.05

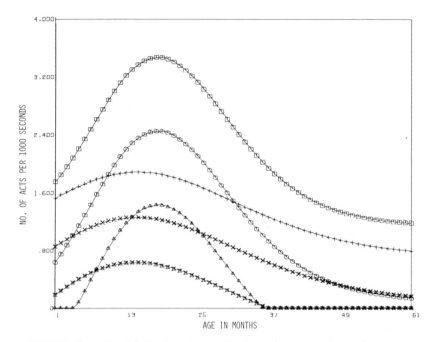

FIGURE 5. Predicted rates of play in vervet monkeys.
☐ = *upper 95% confidence limit in males; o = predicted male*
mean; Δ = lower 95% confidence limit in males; + = upper 95%
confidence limit in females; x = predicted female mean; ⋈ =
lower 95% confidence limit in females.

the computations, including the predicted mean rate of play
and the 95% confidence limits of the predicted mean. Lower
limits of the predicted mean rate are plotted as zeroes when
their calculated values are negative numbers.
    Residuals are examined statistically for consistency of
variance by computing the Spearman rank correlation coefficient (adjusted for ties) between the value of the residual
and the independent variable, age (male r_s = -.0136; p > .05,
and female r_s = -.1042; p > .05). A probability plot of standardized residuals verifies normality of the error terms, lack
of important outlying data points, independence of residuals
with case order, and linearity of residuals. A regression
line fitted through the residuals has an estimated slope of
0.00027 in males and 0.00001 in females. This examination of
residuals does not detect any systematic departure from the
aptness of the model. Furthermore, such small values for lack
of fit F tests support the conclusion that a third degree
polynomial function is appropriate.

Conversion to the proportion of area under the curve verifies observations that have already been made. That is, that
males perform a much larger proportion of their total play
activities between the ages of 9 and 34 months ($Z = 3.87$;
$p < .01$). Likewise, female adults contribute a much greater
proportion of play scores to the total female play activities
than do male adults to total male play ($Z = 4.99$; $p < .01$).

IV. DISCUSSION AND CONCLUSIONS

This chapter began with a reference to the consistency
with which observers seem to be able to recognize play. Play,
in this context, is a subjective category (an English noun)
that is difficult to define. Indeed, the American College
Dictionary (Barnhart and Stein, 1957) lists 55 definitions.
Perhaps the breadth of this noun's interpretation contributes
to the apparent observer agreement. There is a large gulf between a dictionary view of play and units that have been used
to describe vervet play. Fedigan's (1971) analysis is perhaps
the most specific treatment from a descriptive perspective.
Some of the problems associated with observing, recording, and
analysis of play behavior and the rationale by which we dealt
with them are discussed below.
    While it may be easy to agree that a play event is occur-
ring, it is not at all conspicuous what the boundaries of that
event might be. For example, it is not always clear to an observer whether all parties involved in a play event are playing. A juvenile of a subordinate mother may in play, pull the
tail of the most dominant female and dash away as she responds
with aggression. Likewise, a bout of play chasing and cuffs
can frequently increase in intensity among subadult and adult
males until it becomes a fight. More commonly, a higher ranking animal may be actively engaged in wrestle and chase with a
subordinate who seems intent primarily on escaping the interaction. Often the notation of the direction of the encounter
is problematic. At best, in an alternating series of social
play events, initiative may pass from animal to animal so
often that after the earliest noted encounter, it becomes inappropriate to judge who is the initiator and who is the recipient of play. One might propose that a detailed examination of the structural elements could clarify the problem of
the boundaries of play. However, Dolan (1975) notes that the
presence of a play-specific signal (in this case, a play face,
play bounce, or play duck among Sykes' monkeys) is associated
with *lower* probabilities of subsequent play than invitations
to play which do not include these play-specific acts.
    Further, it is difficult to determine all participants

involved in a play encounter. Solitary play in a group which
contains so many potential playmates is not a usual expression
of play in vervet monkeys. This is particularly true of object play. Objects were introduced into the cage and left
until they were demolished. Competition among animals for access to new or novel objects was intense and frequent enough
that observers eventually incorporated *take object from* in the
ethogram as a behavior pattern that readily reflects dominance
status among juveniles. Consequently, the decision was made
to combine all types of play into a single category, social
play. It would, perhaps, be useful to distinguish between interactive and parallel play (Norikoshi, 1974), but we chose
not to do so at this time in order to avoid the uncertain distinction between parallel, simultaneous, and visually coordinated play activities.

   An example of object play may clarify this point. One set
of objects that remained in the cage for a long period included small tin water basins from the shipping crates. Several juveniles began wearing basins on their heads like toy
military helmets. A juvenile engaged in this activity had the
attention, if not the following, of other juveniles who were
contesting for possession of the play object. The basin would
frequently slide forward to cover a wearer's eyes, resulting
in that juvenile running into objects and sometimes inappropriate adults who responded with distinctly nonplay behavior.
It was easy to note which animal had the basin, but the distinction between play and attention among other participants
was problematic.

   In view of these problems in accurately detecting boundaries of play events, the most conservative approach was to use
the most narrow bounds. In the previous example, only the individual with the object (water basin) was scored as playing.
In addition, all categories of play, including both director
and recipient, are combined into a single category which represents participation (willing or otherwise) in play. The
assumption is that those who are truly initiators and targets
of play will show higher scores for involvement than animals
who are inadvertently involved. This assumption is encouraged
(but not tested) by an examination of correlations for play
behaviors. Spearman correlation coefficients (r_s) are 0.71
for male direct-receive (n = 261 paired scores; p < .001) and
0.77 for females (n = 207 paired scores; p < .001). For each
month, animals with higher scores for directing play also have
high scores for receiving play.

   The type of record that is made in this documentation of
play presents a sampling bias that should be noted. Individuals and age classes whose play is characterized by an alterna-
tion of play with other activities will be more frequently
represented than those whose play is continuous and uninter-

rupted. Subjectively, this appears unimportant, since those
age/sex classes exhibiting the long play sequences seem to be
the same classes who play most often and we would expect similar results had we utilized a durational, rather than a frequency analysis.

Traditional development "stages" are perhaps as much a
matter of convenience for the investigator as they are descriptions of behavioral events. Stages intergrade into each
other and are characterized more by continuity of development
than by discrete elements. As Hinde (1971) points out, stage
descriptions may combine references to both physical and social development and are often used to substitute for an age
label (Infant Two vs. six months old). When one focuses on a
single class of acts, such as play, a regression description
is more useful for comparison to other studies. Even Harlow
and Harlow's (1965) excellent descriptions of rhesus developmental stages give little basis for direct comparison with
subsequent studies.

Regression models are as applicable to behavioral development as they have been to the study of physical growth. Analysis of residuals provides a means of examining the appropriateness of the proposed model. In this study, the curve of the
form expressed in Eq. (2) provides a fairly good description
of the acquisition and subsequent performance of play activities in vervet monkeys under five years of age. It is a mathematical description, but some of its properties support hypotheses that have biological meaning. For example, the coefficient of multiple determination (adjusted R square) suggests
that between 60% and 70% of the variance in play behaviors
follows a predictable relationship with age of the subject.
This is compatible with the hypothesis that the motivation for
play involvement is under strong phylogenetic control.

There can be no doubt that much learning, experience, and
environmental stimulation is involved in producing a rate of
play for an individual. Individual variability is quite large
and must surely reflect history, status, experience, and circumstances which confront an individual. To the extent that
experience is age-dependent, it contributes to a correlation
between age and play activities. Underlying variability is a
recognizable pattern of motivation. If this motivational system is age-dependent and reflects growth and maturational processes, then all vervets should exhibit a basically similar
developmental pattern. Predicted mean values generated by
this regression analysis are a rudimentary description of this
pattern.

The first three months of life exhibit marked acceleration
in rates of play. The apparent strategy is to pass from a
neonate stage of development to a juvenile stage as rapidly
as an infant's physical maturation will permit. A juvenile

vervet monkey is characterized by an extremely high rate of
play behavior. Only in juveniles is the motivation and stimulus for play so intensive that zero scores (nonoccurrence of
play during the sampling for that month) do not occur. There
is a marked decrease in play behaviors as animals become
adults that is independent of individual role performance.
Mean age of menarche among vervets in this group is approximately 30 months and mean age for the birth of a first infant
(excluding one female who never conceived) is approximately 47
months (Bramblett, Pejaver and Drickman, 1975). Even nulliparous females exhibit the same decrease in play behaviors as
they make the transition from the juvenile years (months 3 to
27) to adults (over 28 months of age). The changes are clear
cut and abrupt. Males exhibit an extended juvenile period
(months 3 to 46). All males are behaviorally adult by 47
months. In fact, each of the four males in this age range de-
feated the previous adult male (or males) and became the highest ranking adult male in the group between ages 56 and 62
months.

   The probability of an act occurring at a specific age,
among the total observations of that act for that set of animals, provides a technique for crude but useful comparisons of
behaviors that is sensitive to the shape of the distribution
across an age range rather than the absolute frequency of the
behavior. Thus, one can test for a common developmental pattern among animal classes or groups who have widely differing
overall frequencies of behaviors. This comparison in the present study emphasizes the rather conspicuous sexual difference.
One cannot argue that this is simply a reflection of maternal
play (mothers playing with their infants) since nulliparous
females follow the same patterns.

   The hypothesis that all vervets should exhibit a basically
similar developmental pattern should be testable with a similar regression approach. Predicted mean values from regression calculations from longitudinal data on other vervets can
be transformed to represent their proportionate contribution
to total play. A direct comparison of these values provides
a test for similarity of developmental pattern that is adjusted for differences in overall rate of an act in each study
group. This is, in effect, a comparison of the shape of plots
of lines generated by the predicted mean rates for each month.

ACKNOWLEDGMENTS

   Numerous primatology students helped. Sharon Bramblett
tabulated data, keypunched, edited, and typed the manuscript.
I would like to thank T. E. Rowell for assistance in estab-

lishing the study group and L. M. Fedigan for stimulating an
interest in play activities.

REFERENCES

Barnhart, C. L., and Stein, J.   "The American College Diction-
    ary".   Random House, New York (1957).
Bramblett, C. A.   Vervet monkeys and their mothers.   *Bull. Am.
    Anthropol. Assn. 3,* 29 (1970).
_____   The subordinate mother: an analysis of supportive be-
    haviors.   *Abstracts of the 70th Annual Meeting of the
    American Anthropological Association,* p. 22 (1971).
_____   Social organization as an expression of role behavior
    among Old World monkeys.   *Primates 14,* 101-112 (1973).
_____   Acquisition of social behavior in male vervet monkeys:
    the first four and one-half years.   *Am. J. phys. Anthro-
    pol. 40,* 131 (1974).
_____   "Patterns of Primate Behavior".   Mayfield Publ. Co.,
    Palo Alto (1976a).
_____   Serum cholesterol response to high dietary cholesterol
    in *Cercopithecus. Am. J. phys. Anthropol. 44,* 167 (1976b).
Bramblett, C. A., Pejaver, L. D., and Drickman, D. J.   Repro-
    duction in captive vervet and Sykes' monkeys.   *J. Mammal.
    56,* 940-946 (1975).
Buckley, J.   Sexual differences in the acquisition of grooming
    behaviors in captive vervet monkeys (*Cercopithecus aethi-
    ops*).   *Am. J. phys. Anthropol. 44,* 168 (1976).
Chickering, L.   Changes in social structure in a captive ver-
    vet (*Cercopithecus aethiops*) monkey group.   *Am. J. phys.
    Anthropol. 44,* 170 (1976).
Dolan, K. J.   "Metacommunication in the Play of a Captive
    Group of Sykes' Monkeys".   M.A. Thesis, University of
    Texas, Austin (1975).
Estrada, A., and Bernal, J.   Social play in captive groups of
    vervet monkeys (*Cercopithecus aethiops*).   Unpublished
    manuscript (1977).
Fedigan, L. M.   "Social and Solitary Play in a Colony of Ver-
    vet Monkeys".   M.A. Thesis, University of Texas, Austin
    (1971).
_____   Social and solitary play in a colony of vervet monkeys.
    Primates 13, 347-364 (1973).
Harlow, H. F., and Harlow, M. K.   The affectional systems, *in*
    "Behavior of Nonhuman Primates", Vol. II (A. M. Schrier,
    H. F. Harlow, and F. Stollnitz, eds.), pp. 287-334. Aca-
    demic Press, New York (1965).
Hinde, R. A.   Development of social behavior, *in* "Behavior of
    Nonhuman Primates", Vol. III (A. M. Schrier, and F. Stoll-

nitz, eds.), pp. 1-68. Academic Press, New York (1971).

Hunter, J. "Effect of Object Size, Character, and Context on Manual Grip in *Cercopithecus*". M.A. Thesis, University of Texas, Austin (1973).

Lancaster, J. B. Playmothering, the relations between juvenile females and young infants among free-ranging vervet monkeys (*Cercopithecus aethiops*). *Folia primatol. 15*, 161-182 (1971).

Loizos, C. Play behavior in higher primates: a review, *in* "Primate Ethology" (D. Morris, ed.), pp. 176-218. Aldine, Chicago (1967).

Mason, W. A. The social development of monkeys and apes, *in* "Primate Behavior: Field Studies of Monkeys and Apes" (I. DeVore, ed.), pp. 514-543. Holt, Rinehart and Winston, New York (1965).

Neter, J., and Wasserman, W. "Applied Linear Statistical Models". Richard D. Irwin, Inc., Homewood (1974).

Norikoshi, K. The development of peer-mate relationships in Japanese macaque infants. *Primates 15*, 39-46 (1974).

Rowell, T. E. Reproductive cycles of two *Cercopithecus* monkeys. *J. Reprod. Fertil. 22*, 321-338 (1970).

_____ Organization of caged groups of *Cercopithecus* monkeys. *Anim. Behav. 19*, 625-645 (1971).

_____ "Social Behaviour of Monkeys". Penguin, Baltimore (1972).

_____ The concept of social dominance. *Behavioral Biol. 11*, 131-154 (1974a).

_____ Contrasting adult male roles in different species of nonhuman primates. *Arch. sex. Behav. 3*, 143-149 (1974b).

Siegel, S. "Nonparametric Statistics for the Behavioral Sciences". McGraw-Hill, New York (1956).

Snedecor, G. W., and Cochran, W. G. "Statistical Methods", 6th Ed. Iowa State University Press, Ames (1967).

Struhsaker, T. T. Behaviour of vervet monkeys, *Cercopithecus aethiops*. *Univ. Calif. Publ. in Zoology 82*, 1-74 (1967).

Ward, J. H., and Jennings, E. "Introduction to Linear Models". Prentice Hall, Englewood Cliffs, New Jersey (1973).

THE BEHAVIOR OF GONADECTOMIZED RHESUS MONKEYS
I. PLAY[1]

James Loy
Kent Loy
Donald Patterson
Geoffrey Keifer
Clinton Conaway

Department of Sociology and Anthropology
University of Rhode Island
Kingston, Rhode Island

I. INTRODUCTION

Between November, 1971, and May, 1976, the Caribbean Primate Research Center was the site of a longitudinal study into the effects of gonadectomy on the behavior of socially-living rhesus monkeys. This study was considered to be exploratory since few previous investigations had examined the behavior of gonadectomized monkeys within the setting of a social group (a recent exception is the study by Cochran and Perachio, 1977). The more frequent design has been to house monkeys singly and pair them only for short test periods (e.g. Michael, Herbert and Welegalla, 1966; Michael and Wilson, 1974). Recognizing the exploratory nature of the project, and not knowing precisely what post-operative behavioral changes might occur, information was collected on a variety of behavior patterns, including friendly interactions (grooming, body contact, sitting in proximity, play), sexual interactions and auto-eroticism, and agonistic exchanges (and dominance). The results of the study will be published as a series of papers, beginning with the present article on play behavior.

[1]During this study, the Caribbean Primate Research Center was supported by Contract NIH-DRR-71-2003 from the National Institutes of Health. Portions of the gonadectomy study were supported by Grant No. GB-44252 from the National Science Foundation to J. Loy.

49

II. METHODS

A. *Animal Acquisition and Group Formation*

In late November, 1971, 33 yearling and two-year-old rhesus monkeys (*Macaca mulatta*) were removed from Cayo Santiago (Puerto Rico) and transferred to the colony near La Parguera, Puerto Rico. Both colonies were facilities of the Caribbean Primate Research Center. The 33 monkeys (17 males and 16 females) had all been born into the same free-ranging social group on Cayo Santiago (Group A) during the birth seasons of 1969 and 1970, and their matrilineal relationships were known (Loy and Loy, 1974). Upon arrival at La Parguera, all 33 animals were placed in a one-half acre corral for initial preoperative observations. An analysis of the dominance relations and other behaviors seen in the juvenile group of 33 monkeys has been published elsewhere (Loy and Loy, 1974).

On 25 January, 1972, the monkeys were divided into an experimental group of 17 animals and a control group of 16 animals (Table I). Monkeys were assigned to either the experimental or control group as follows:

(1) The groups were matched for ages and sexes as nearly as possible. Each group contained six monkeys born in 1969 (three males and three females) and either 10 or 11 monkeys born in 1970.

(2) The groups were matched for average dominance rank of their members.

(3) All 10 sibling pairs were split, with one member going to each group.

(4) Monkeys who had displayed adult sexual behavior or cycling during the initial observations were assigned to the control group when possible.

Following group assignment, the 17 experimental monkeys were relocated in a one-quarter acre outdoor corral and observations on both groups were resumed prior to surgery. The one-quarter acre corrals used in this study were constructed of sheet metal and chain link fencing, and allowed visual, auditory, olfactory and limited tactile communication with the free-ranging rhesus population of the La Parguera colony. Each corral was equipped with a set of metal "trees", three sheet metal shelters for shade and climbing, water dispensers and a feeder. Throughout the study, Wayne Monkey Diet and water were available to the animals *ad libitum*, and fruits and rice were occasionally given as dietary supplements.

TABLE I. Composition of Study Groups[a]

		Controls		Experimentals	
	Name	Approximate Dominance[b]	Name	Approximate Dominance	
1969 Births	*Females*		*Females*		
	277	5	260	4	
	242	10	272	6	
	235	11	273	9	
	Males		*Males*		
	266	1	246	2	
	230	7	241	12	
	387	3	231	8	
	$\bar{X}$ dominance = 6.1		$\bar{X}$ dominance = 6.8		
1970 Births	*Females*		*Females*		
	400	5	367	15	
	333	14	370	1	
	331	12	366	7	
	379	20	386	11	
	388	19	354	16	
	Males		*Males*		
	377	8	355	6	
	385	3	330	17	
	399	10	389	9	
	334	2	371	4	
	383	20	335	20	
			301	18	
	$\bar{X}$ dominance = 11.3		$\bar{X}$ dominance = 12.1		

[a]The following sibling pairs were split: 277-355; 242-367; 235-330; 266-370; 230-389; 260-400; 272-377; 273-333; 246-385; 231-331.

[b]'Within age-class' dominance ranks as determined during preoperative study (Loy and Loy, 1974).

B. Gonadectomies

Between 23 March - 7 April, 1972, all of the experimental monkeys (with one exception to be noted below) were bilater-

ally gonadectomized (Table II). These operations were performed pre-pubertally for all males, and peri-pubertally for
all females (i.e., some experimental females had exhibited
some sex skin swelling and coloration prior to ovariectomy).
No surgical manipulations of any sort were performed on the
control monkeys at this time.

The exceptional experimental monkey just mentioned was female 370, who proved to be extremely difficult to ovariectom-
ize. On 23 March, 1972, female 370's right ovary was removed
and a search failed to locate the left ovary. A second lap-
arotomy was done on 11 May, 1972, and 370's left ovary was
found and removed. The second operation only temporarily disrupted 370's sexual cycling, however, and during August, 1972,
she showed the return of sex skin swelling and coloration. On
11 September, 1972, a third operation on 370 revealed no tissue that was clearly ovarian. Even after this surgery, however, 370 continued to show menstrual and color cycles for the
duration of the study. Female 370 was not removed from the
experimental group despite her failure to respond satisfactorily to gonadectomy. She was the highest ranking female in the
group and it was felt that her removal would introduce considerable social disruption. In this paper, play interactions
of 370 are included with those from the clearly ovariectomized
females.

Post-operative observations on the gonadectomized monkeys
continued without significant interruption from April, 1972,
through May, 1976. The only additional manipulations of the
experimental animals that might have been at all important
were second laparotomies for four of the ovariectomized females
(367, 366, 386, 273), who were showing increases in sex skin
color two years after their initial operations. All four females were surgically checked in May or June, 1974, and all
yielded small amounts of (possible) ovarian tissue that apparently had developed since their ovariectomies.

C. Controls

The one-quarter acre corral for the control monkeys was not
completed until May, 1972. Between 25 January and 10 May,
1972, the controls remained in the original one-quarter acre
corral, and on 11 May, they were moved to their new enclosure.
Pre-operative observations continued on the control monkeys
during January - March, 1972.

The social structure of the control group changed during
1973-1974 due to successful breeding. Four control females
conceived during the mating season of 1972 and gave birth between April - June, 1973. The four infants (two males and two
females) were left in the control group until 4 February, 1974,

Table II. Histories of Study Groups

Date	Event
19 November 1972	The first of 33 rhesus juveniles arrived at La Parguera from Cayo Santiago. Beginning of study.
25 January 1972	Formation of Control and Experimental groups. Experimentals moved to 1/4-acre corral #5.
23 March 1972	Bilateral castrations - males 231, 246, 241; bilateral ovareectomies - females 273, 272, 260, 367; right ovary removed - female 370.
24 March 1972	Bilateral ovariectomies - females 354, 386, 366.
7 April 1972	Bilateral castrations - males 389, 335, 330, 301, 371, 355.
11 May 1972	Left ovary removed - female 370. On same date, controls moved to 1/4-acre corral #6.
11 September 1972	Female 370 laparotomized; no positive ovarian tissue found; stub of right fallopian tube removed.
19 October 1972	Castrate male 389 died.
October - December 1972	Four control females conceived.
April - June 1973	Four infants born into control group.
31 May 1973	Control female 379 died.
October 1973 - February 1974	Six control females conceived.
8 December 1973	Ovariectomized female 354 died.
4 February 1974	All 1973 infants removed from control group.
April - May 1974	1974 control infants removed from mothers immediately after birth.

Continued on following page

Date	Event
30 May 1974	All control males bilaterally vasectomized.
31 May 1974	Female 367 laparotomized; "probable" ovarian nodules removed both sides.
19 June 1974	Females 366, 273 and 386 all laparotomized; all produced "possible" ovarian tissue. On same date, female 400 laparotomized for removal of fetus.
22 July 1974	Ovariectomized female 366 died.
21 May 1976	Final day of observations. End of study.

when all were permanently removed in order to reestablish matching structures between the control and experimental groups. During the 1973 mating season, six control females conceived. As the 1974 infant crop was born, the infants were immediately taken away from their mothers. (One control female, #400, had her 3-4 month fetus surgically removed.) Thus, two sets of pregnancies and one year with infants were allowed to occur within the control group. On 30 May, 1974, all eight of the control males were bilaterally vasectomized to prevent further pregnancies. The control males were returned to their corral immediately after post-operative recovery, and no significant interruption in observations occurred. During the final two years of the study, no conceptions or births occurred within the control group. Major manipulations and important events in the histories of both groups are listed in Table II.

D. Sampling Procedures

Post-operative observations on both groups were conducted from April, 1972, through May, 1976. Several observation techniques were used during the study, including (a) random sampling of interactions across an entire social group ("field note" sampling), (b) focal sampling of particular behavior patterns for individual monkeys, and (c) "scan" sampling across a social group. All of the data in the present paper were collected using the "scan" sampling procedure, which will be described in more detail.

The scan sampling technique was designed to produce accurate information on the frequency of behaviors considered important. During each 15-minute scan sample, a social group was closely monitored for all instances of fighting, social grooming, sitting in body contact, playing and mounting. As these behaviors were observed, they were tic-marked on a sample sheet according to the sexes of the participants. The identities of the interacting monkeys were not recorded. The scan samples yielded accurate estimates of the frequencies of key behaviors, as well as data on the participation rates of the two sexes in these interactions. Each group was typically scan sampled twice between 0600-1200 hours on Monday through Friday. An attempt was made to collect a total of twenty-five 15-minute samples per group each month.

Scan sampling was begun in July, 1972, three months after the gonadectomies, and continued for 47 months, ending in May, 1976. A total of 2031 scan samples were collected (507.75 hours).

E. Play Defined

As noted above, one of the behavior patterns recorded during scan samples was play. Only social play was recorded; solitary play, with or without object manipulation, was ignored. In addition, in order for an interaction to be scored as play, both of the animals involved had to be clearly participating in playful behaviors. Interactions in which play overtures by one monkey were ignored or responded to in a nonplayful manner by a potential partner were not scored as play bouts. Play was thus defined as a reciprocal, social interaction pattern.

No attempt was made to differentiate between the several varieties of social play exhibited by rhesus monkeys (Hinde and Spencer-Booth, 1967). Play interactions consisted of one or more of the following motor patterns: wrestling, chasing, slapping, mouthing, pulling, pushing, running away, and jumping on or away from another monkey.

F. Calculation of Play Indices

The play behavior seen within each social group was measured through the use of seven indices, which were calculated each month from scan sampling data.

1. *Play/Dyad/Hour.* A group's dyadic play episodes for all monkeys were totalled for the month. This total was then divided by the total dyads in the group:

$$\left(\frac{N \times N-1}{2}\right)$$

and the resulting figure was further divided by the total scan sample hours for the month.

2. Male-Male Play/Dyad/Hour; 3. Female-Female Play/Dyad/ Hour; 4. Isosexual Play/Dyad/Hour; 5. Heterosexual Play/Dyad/ Hour. The calculations for indices 2-5 followed a common format. In each case, the month's total of dyadic play episodes for the category under consideration (e.g., male-male, heterosexual) was divided by the total number of dyads of that category. The resulting figure was then further divided by the group's total scan sample hours for the month.

6. Play Episodes/Male/Hour. All episodes of male play were totalled for the month. (A play interaction between two males contributed two play episodes to the month's total, whereas a play interaction between a male and a female contributed a single play episode.) The month's play total was then divided by the number of males in the group, and the resulting figure was further divided by the scan sample hours for the month.

7. Play Episodes/Female/Hour. Calculations were the same as for Index #6, only for females.

It should be noted that some play interactions occurred between control group adults and infants between May, 1973, and January, 1974. These interactions were not used in the calculation of any control group indices with the exceptions of play/male/hour and play/female/hour (Indices #6 and #7).

The monthly values on these various indices were used as the basic data for inter-group and intra-group play comparisons. Inter-group behavioral comparisons were performed with one-way analyses of variance. Intra-group tests (e.g., for seasonal differences) were done with the "t" test for related populations. In all cases, the level of significance (alpha) was set at .05.

For some comparisons, play was grouped by post-operative year (POY). Post-operative year 1 began in April, 1972, and extended through March, 1973. However, since scan sampling did not begin until July, 1972, POY 1 is represented by only nine months of data. Post-operative year 2 has a full 12 months of data, while POY 3 has only 11 months of data due to the loss of the August, 1974, scan samples in the mail. Post-operative year 4 contains 14 months of data since the two extra months at the end of the study (April and May, 1976) were included in that year for our calculations.

For other behavioral comparisons, play was grouped according to whether it occurred during the mating season or the non-mating season. The seasonal terms refer to the presence or absence of sexual behavior within the control group. The control mating season occurred in the fall and winter months, and was marked by heterosexual copulations and females with vaginal plugs of ejaculate (Loy, 1971). During the spring - summer non-mating season, copulatory behavior was not seen and females were not observed with vaginal plugs.

G. Hormonal Analysis

One of the main objectives of the study was to determine how gonadal hormone levels correlated with behavior among both the control and gonadectomized monkeys. To that end, blood samples were regularly collected for hormone assay. Between May, 1973, and May, 1976, all monkeys were caught and bled each year during the (control) mating season and again during the non-mating season. Whole blood was centrifuged to produce plasma and the plasma samples were frozen prior to shipment to the Wisconsin Regional Primate Research Center for assay. Plasma samples from females were assayed for estradiol and progesterone, while samples from males were assayed for testosterone.

III. RESULTS

A. Assays

The assay results for the ovariectomized females are shown in Table III. (The January, 1976, and May, 1976, samples had not been analyzed at the time of writing.) The estradiol and progesterone values for all females (except 370) were uniformly very low. Only female 370 showed hormone fluctuations typical of an intact female (Hotchkiss, Atkinson and Knobil, 1971; Niswender and Spies, 1973). The assay results for the control females were of little value for direct comparisons with the ovariectomized monkeys, since the hormone levels in the former animals depended on their menstrual cycle phase and whether or not they were pregnant when bled.

The control males showed a clear seasonal cycle in testosterone values (Table IV). Testosterone levels were highest in the mating season and lowest during the non-mating season. These results agree with the data reported by Gordon, Rose and Bernstein (1976). The castrated males showed no seasonal changes in their very low testosterone values (Table IV). The

Table III. Estradiol and Progesterone Values for Ovariectomized Females

Date of Bleeding	Estradiol			Progesterone			Female 370		
	No. of Females	Mean Age (mos)	Mean Estradiol[a]	No. of Females	Mean Age (mos)	Mean Progesterone[b]	Age	Estradiol	Progesterone
30 May '73	6	42.6	12.2	7	43.4	0.68	39.5	16.5	0.43
27 Dec '73	6	51.2	13.2	6	51.2	0.49	46.5	226.5	0.56
31 May '74	5	55.3	9.5	6	56.2	0.36	51.5	72.5	1.19
9 Jan '75	5	64.7	10.1	5	64.7	0.36	58.5	63.0	5.03
5 June '75	5	69.7	13.5	5	69.7	0.31	63.5	26.0	0.49

[a] Estradiol and progesterone values are means for all ovariectomized females except #370. Estradiol values are in pg/ml plasma.

[b] Progesterone levels are in ng/ml plasma.

Table IV. Testosterone Values for Control and Castrated Males

Date of Bleeding	Season[a]	Controls			Castrates		
		No. of Males	Mean Age (Mos)	Mean Testosterone[b]	No. of Males	Mean Age (Mos)	Mean Testosterone
30 May '73	NM	8	42.7	0.38	8	43.4	0.20
27 Dec '73	M	8	49.7	2.82	8	50.4	0.13
30/31 May '74	NM	8	54.7	0.80	8	55.4	0.12
9/10 Jan '75	M	8	61.7	2.93	8	62.4	0.15
4/5 June '75	NM	8	66.7	1.84	8	67.4	0.18

[a] Mating (M) or non-mating (NM) season in control group.

[b] Testosterone values given in ng/ml plasma.

testosterone values of our castrated males compare well with the levels reported for rhesus male castrates by Resko and Phoenix (1972).

B. *Pre-operative Play*

Loy and Loy (1974) presented some information on the play behavior seen during the pre-operative observations on the juvenile group of 33 monkeys. However, that paper did not include information on the play frequencies of monkeys destined to become controls as compared to those destined to be gonadectomized. In an attempt to determine whether unusually playful monkeys had been assigned to one group or the other, separate analyses of variance were calculated comparing (1) the pre-control females with the pre-ovariectomized females for the frequency of play/female/hour, and (2) the pre-control males with the pre-castrated males for the frequency of play/ male/hour. Pre-control females averaged 0.13 play episodes/ female/hour as compared to 0.12 episodes/female/hour for the pre-ovariectomized animals ($F = 0.00$, df $= 1,14$, $p > .05$). Pre-control males averaged 0.36 play episodes/male/hour as compared to 0.44 play episodes/male/hour for males destined to be castrated ($F = 0.38$, df $= 1,15$, $p > .05$). It was concluded that the two groups began the study with equal potential for playful behavior.

C. *Post-operative Play: Annual and Seasonal Results*

1. *Play/Dyad/Hour*. A general measure of the total amount of play occurring within a social group was provided by the index play/dyad/hour. When play/dyad/hour was grouped according to post-operative year, it was found that there was significantly more play among the gonadectomized monkeys than among the controls in all four years (POY 1, $F = 5.67$, df $= 1$, 16, $p < .05$; POY 2, $F = 19.70$, df $= 1,22$, $p < .01$; POY 3, $F = 18.00$, df $= 1,20$, $p < .01$; POY 4, $F = 31.81$, df $= 1,26$, $p < .01$; Figure 1).

Similarly, the gonadectomized monkeys showed more play/ dyad/hour than the controls during the post-operative mating seasons ($F = 7.53$, df $= 1,50$, $p < .01$) and non-mating seasons ($F = 4.39$, df $= 1,38$, $p < .05$). Neither group showed significant intra-group changes in play/dyad/hour from the total mating seasons to the total non-mating seasons.

While these data are interesting, they do not provide information about the relative contributions of males and females to the groups' play totals. Other indices were required for the calculation of sex-class play.

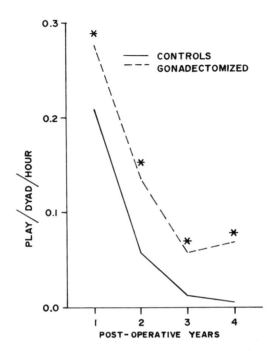

FIGURE 1. Play/Dyad/Hour. An asterisk indicates a significant inter-group difference for the post-operative year shown.

2. *Play/Female/Hour.* Analyses of the frequency of play/female/hour were done for each group across the four post-operative years. Figure 2 shows the relationship between play by the control females and the subjects' mean age. A clear inverse relationship was found, with play decreasing rapidly as the control females grew older. After significant decreases during the first two post-operative years (POY 1 vs. POY 2, t = 2.53, p < .025; POY 2 vs. POY 3, t = 4.04, p < .005), play by the control females essentially reached zero during post-operative year 3, when the mean female age ranged from 53.4 – 65.4 months.

Similarly, the ovariectomized females showed significant drops in play/female/hour as they grew older (POY 1 vs. POY 2, t = 3.32, p < .01; POY 2 vs. POY 3, t = 4.08, p < .005; POY 3 vs. POY 4, t = 1.82, p < .05; Figure 3). It should be noted that in both groups, females played much less frequently than males. (Compare the rates of occurrence in Figures 2 and 3 with those shown for males in Figures 5 and 6.)

An annual comparison revealed a significant difference between the two female populations only during post-operative

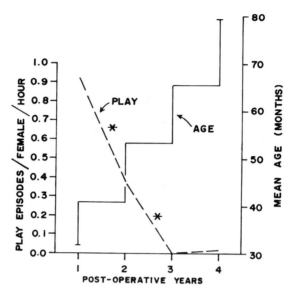

FIGURE 2.   Play/Female/Hour and Age: Control females.   An asterisk indicates a significant intra-group change in play frequency between successive post-operative years.

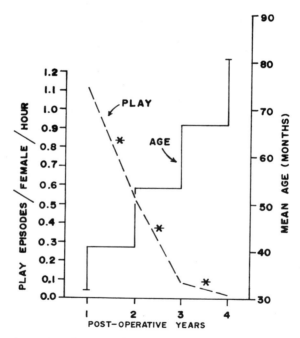

FIGURE 3.   Play/Female/Hour and Age: Ovariectomized fe-males.   See Figure 2 for meaning of asterisk.

year 3 (F = 7.27, df = 1,20, p < .05), with the ovariectomized females playing slightly more frequently than the controls (Figure 4). The significant difference during year 3 does not reflect a large difference in actual play between the two groups of females, however. As control female play essentially ceased, a play episode involving an ovariectomized female was occurring about once in every 2 hours of observation. A seasonal comparison revealed non-significant inter-group differences in play/female/hour across the total post-operative mating seasons and non-mating seasons.

No significant intra-group differences in play/female/hour were found between the total mating seasons and the total non-mating seasons for either female population.

Overall, it appears that there was little difference in the frequency of play/female/hour in the two groups, and that in both female populations, play frequency decreased rapidly with increasing age.

3. *Play/Male/Hour*. The frequencies of play/male/hour for the control males during the four post-operative years are shown in Figure 5, along with the concurrent mean male ages.

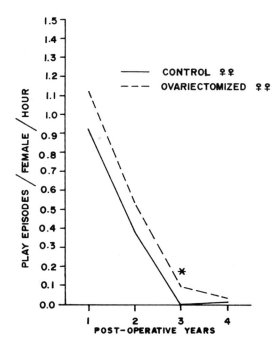

FIGURE 4. *Play/Female/Hour: Controls versus Ovariectomized. Asterisk indicates significant inter-group difference for post-operative year.*

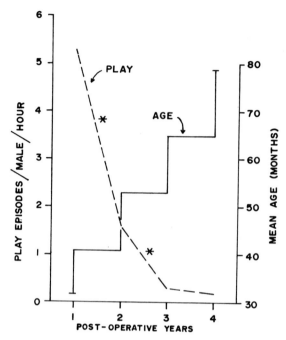

FIGURE 5. Play/Male/Hour and Age: Control males. Asterisk indicates significant intra-group change in play frequency between successive post-operative years.

Control male play frequency was clearly related to age and dropped significantly as the males grew older (POY 1 vs. POY 2, t = 5.32, p < .0005; POY 2 vs. POY 3, t = 4.06, p < .005). During the third and fourth post-operative years, the control males showed infrequent play.

Similarly, the castrated males showed significant drops in play/male/hour as age increased (POY 1 vs. POY 2, t = 8.22, p < .0005; POY 2 vs. POY 3, t = 9.04, p < .0005; Figure 6).

While both male populations showed age effects on male play frequency, a comparison of the two groups also revealed treatment-related differences in male play. The castrated males showed significantly more play/male/hour than the control males during every post-operative year (POY 1, F = 5.40, df = 1,16, p < .05; POY 2, F = 19.11, df = 1,22, p < .01; POY 3, F = 16.94, df = 1,20, p < .01; POY 4, F = 29.89, df = 1,26, p < .01; Figure 7). Furthermore, while the control males leveled off at means of 0.34 and 0.19 play episodes/male/hour during years 3 and 4, respectively, the castrated males showed 1.36 play episodes/male/hour during the third

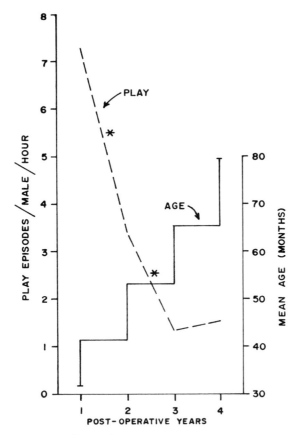

FIGURE 6. Play/Male/Hour and Age: Castrated males. See Figure 5 for meaning of asterisk.

post-operative year and an even higher value, 1.55 episodes/ male/hour, during the fourth year.

Neither the control males nor the castrated males showed significant intra-group differences in play/male/hour between the total mating seasons and the non-mating seasons. When inter-group comparisons were made on a seasonal basis, it was found that the castrated males showed significantly more play/ male/hour than the controls during the combined mating seasons ($F = 6.27$, $df = 1,50$, $p < .05$), but not during the combined non-mating seasons.

4. *Isosexual Play.* During all four post-operative years, the gonadectomized monkeys showed significantly higher levels of isosexual play/dyad/hour than the control monkeys (POY 1, $F = 5.42$, $df = 1,16$, $p < .05$; POY 2, $F = 21.75$, $df = 1,22$,

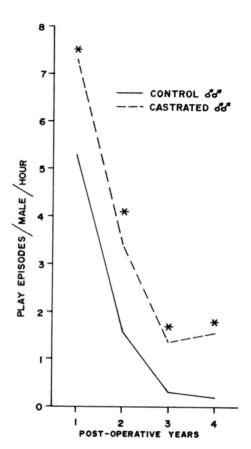

FIGURE 7.   *Play/Male/Hour: Controls versus castrates.*
Asterisk indicates significant inter-group difference for
post-operative year.

p < .01; POY 3, F = 14.65, df = 1,20, p < .01; POY 4, F =
29.38, df = 1,26, p < .01; Figure 8).   These differences were
related to the extremely frequent play of the castrated males,
and to the fact that most play in both groups was between like-
sexed monkeys.   The data on male-male play/dyad/hour show much
higher levels among the castrates than among the controls,
with significant differences during post-operative years 2-4
(POY 2, F = 22.35, df = 1,22, p < .01; POY 3, F = 15.43, df =
1,20, p < .01; POY 4, F = 30.73, df = 1,26, p < .01; Figure 9).
In contrast, the frequencies of female-female play/dyad/hour
did not differ significantly between the control and ovariectomized females during any post-operative year (Figure 10).
     The castrated males showed significantly more male-male
play/dyad/hour than the control males during the total mating

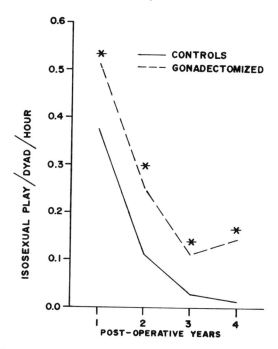

FIGURE 8. Isosexual Play/Dyad/Hour. Asterisk indicates
significant inter-group difference for post-operative year.

seasons (F = 6.64, df = 1,50, p < .05), but not during the
total non-mating seasons. Neither population of males showed
significant intra-group seasonal changes in male-male play.
The two female populations showed non-significant inter-group
differences in the frequency of female-female play during both
the mating seasons and the non-mating seasons. Similarly, the
intra-group seasonal changes in female-female play were not
significant for either female group.

5. Heterosexual Play. The occurrence of heterosexual
play was low in both groups. Even during post-operative year
1, when both groups were most playful, heterosexual play occurred only 1/6-1/7 times as frequently as isosexual play.
Clearly, like-sexed play partners were preferred in both
groups.

The gonadectomized monkeys showed significantly more heterosexual play/dyad/hour than the controls during post-operative years 2 and 3 (POY 2, F = 6.33, df = 1,22, p < .05;
POY 3, F = 5.71, df = 1,20, p < .05; Figure 11). These differences are probably related to the slightly more frequent
play of the ovariectomized females as compared to the control

females, and to the fact that the very playful castrated males
would occasionally include females in their play interactions.

Inter-group differences in the frequency of heterosexual
play were non-significant during the total mating seasons, as
well as during the total non-mating seasons. In addition,
neither the controls nor the gonadectomized monkeys showed
significant intra-group changes in the frequency of heterosexual play between the total mating seasons and the total non-
mating seasons. We had expected that heterosexual play among
the controls might increase during the mating season when the
monkeys experience a shift toward heterosexual orientation in
other activities such as mounting and grooming (J. Loy *et al.*,
unpublished data). The lack of seasonal changes in heterosexual play among the controls probably reflects the fact that

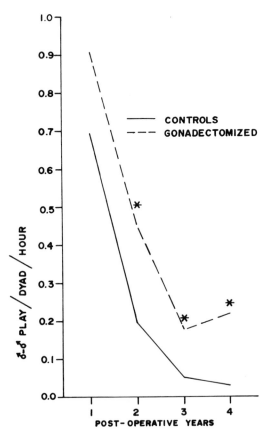

*FIGURE 9. Male-Male Play/Dyad/Hour. Asterisk indicates
significant inter-group difference for post-operative year.*

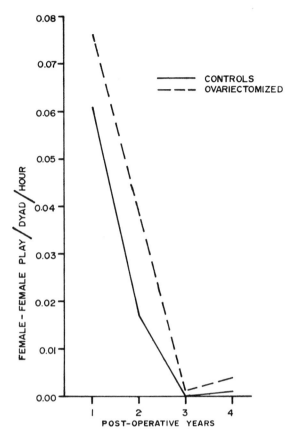

FIGURE 10. Female-Female Play/Dyad/Hour. Asterisk indicates significant inter-group difference for post-operative year.

play is so rapidly lost from the typical behavior of intact females that heterosexual play becomes a rare occurrence at any time of the year.

D. Total Post-operative Play

In addition to the annual and seasonal comparisons, the play data were summed across the entire 47-month period of scan sampling for overall inter-group comparisons (Table V). These comparisons revealed that the gonadectomized monkeys showed significantly more play than the controls in four categories: play/dyad/hour, male-male play/dyad/hour, isosexual play/dyad/hour and play/male/hour. Non-significant differ-

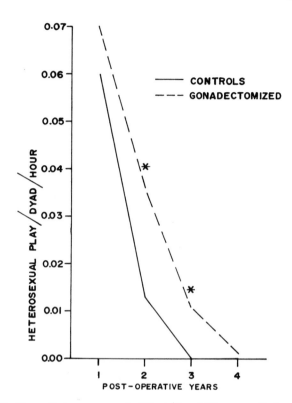

FIGURE 11.   Heterosexual Play/Dyad/Hour.   Asterisk indicates significant inter-group difference for post-operative
year.

ences between the two groups were found for female-female
play/dyad/hour, heterosexual play/dyad/hour and play/female/
hour.

Two key variables, play/male/hour and play/female/hour,
appear to explain the group differences seen in Table V.   The
fact that across the entire post-operative period, the castrated males were showing almost twice as much play/male/hour as
the control males led to significant inter-group differences
in the three other play categories which were most heavily influenced by male participation.   Similarly, the fact that the
control and ovariectomized females showed very similar overall
play frequencies tended to reduce to non-significant levels
the inter-group differences in the play categories which were
most heavily influenced by the presence or absence of female
participation.

Table V. Total Post-Operative Play: Inter-Group Comparisons

Index	Mean Value Controls	Mean Value Gonadectomized	F	df	p Values
Play/dyad/hour	0.06	0.12	12.11	1,90	p < .01
Male-male play/dyad/hour	0.20	0.40	10.60	1,90	p < .01
Isosexual play/dyad/hour	0.12	0.24	13.27	1,90	p < .01
Play/male/hour	1.58	3.11	9.88	1,90	p < .01
Female-female play/dyad/hour	0.02	0.03	1.55	1,90	p > .05
Heterosexual play/dyad/hour	0.02	0.03	3.08	1,90	p > .05
Play/female/hour	0.28	0.39	1.17	1,90	p > .05

IV. DISCUSSION

Overall, the occurrence of playful behavior appears to
follow a fairly simple pattern among intact rhesus monkeys,
and a pattern that is similar for both sexes. Both females
and males show frequent play during their infantile and juvenile years (Hinde and Spencer-Booth, 1967; Loy and Loy, 1974;
Figures 2 and 5, this study), but after the early peak, play
declines with increasing age and becomes a rare occurrence
among physically adult animals. A sociobiological explanation
of the ultimate or evolutionary cause of this pattern of play
behavior points out its adaptive characteristics (Barash,
1977). The reproductive success of both males and females
should be enhanced by frequent play while still immature, following the theories that play increases knowledge of the environment, provides valuable practice in motor and communication skills, aids in the development of social relationships
among peers, and generally furthers the social integration of
the maturing monkey (Dolhinow and Bishop, 1970; Loizos, 1967).
Neither sex should suffer from the devotion of a great deal of
time and energy to pre-pubertal play, since that time and energy could not be directed toward adult reproduction, anyway.
Following sexual maturity, however, sociobiological theory
might predict that further play, by either males or females,
generally will be non-adaptive, and that the monkeys' atten-
tions and efforts will be turned toward other activities. An
adult male might best increase his reproductive fitness by
concentrating on agonistic behavior with other males as he
achieves and maintains a spot in the male dominance hierarchy.
Several studies suggest that among cercopithecoid monkeys,
high male dominance rank is correlated with reproductive success (Conaway and Koford, 1965; Hall and DeVore, 1965; Loy,
1971), although the evidence is somewhat equivocal (Duvall,
Gordon and Bernstein, 1976; Hausfater, 1975). Adult males
also increase their fitness by directing their time and energy
toward courting and copulating with receptive females, as well
as defending the group's offspring against external threats.
Adult females increase their individual fitness by competing agonistically with peers as they move into and maintain a
position in the female dominance hierarchy. Among rhesus monkeys, a female's dominance rank affects her reproductive success not only by influencing her access to important environmental resources, but also by strongly affecting the social
development of her offspring (Koford, 1963; Loy, 1975; Loy and
Loy, 1974; Sade, 1967). In addition, adult females increase
their fitness by spending time grooming and being groomed by
the members of their social support unit - the matrilineal
family - and by devoting time and energy to courting and

copulating with adult males. Finally, an adult female must take on the responsibilities for nursing and infant care after the birth of her offspring.

Overall, sociobiological theory would probably predict play to be an adaptive and valuable behavior for immature rhesus monkeys, but a generally non-adaptive behavior for adults. (A minor exception would be the infrequent play that occurs between infants and adults of both sexes. Such play would be beneficial to adults, since it would contribute to the social development of their offspring.) Therefore, assuming that play is, to some extent, genetically based, monkeys who showed frequent play pre-pubertally, followed by infrequent play post-pubertally, should have experienced differential reproductive success and, thus, that ontogenetic pattern of play would have become established in the species.

While the sociobiological explanation of the ultimate causes of rhesus monkey play applies to both sexes, an investigation of the proximate factors affecting play reveals that male play is somewhat more complex than that of females in regard to the variables which support or influence its occurrence. The ontogenetic pattern of play frequency shown by intact rhesus females is primarily a function of age. The present study has shown that pre- or peri-pubertal ovariectomy did not produce significant differences in either the frequency of female play or the pattern of play partner selection when compared with the play behavior of intact control females, all of whom were experiencing regular menstrual/estrous cycles and some of whom were conceiving and bearing young. In both female populations, play occurred much less frequently than did play by males. It appears that the play of rhesus females is not significantly affected by the presence or absence of post-natal stimulation by the gonadal hormones; rather, regardless of fluctuations in her gonadal hormones (and probably regardless of her reproductive success or failure), a rhesus female tends to show a peak in play frequency as an infant or young juvenile (Hinde and Spencer-Booth, 1967; Loy and Loy, 1974) and then to show a decrease in play with increasing age.

In contrast, there seem to be several variables which exert proximate influence on the pattern and frequency of male play behavior. The inter-group comparisons from the present study show that pre-pubertal castration affects the frequency of male play and the decrease of play behavior with increasing age. These behavioral effects appear to be complexly linked to the hormonal condition, growth rate and social relationships of the castrated males.

Testosterone has been linked to several aspects of rhesus male behavior, including aggression (Gordon et al., 1976; Rose et al., 1974) and sexual behavior (Michael and Wilson, 1974; Phoenix, 1974a). Rhesus males typically experience testicular

descent at 3 - 3-1/2 years of age (Conaway and Sade, 1965;
Sade, 1964), and at about this same age, they begin to produce
adult levels of testosterone (Resko, 1967). Also at 3 - 3-1/2
years of age, intact rhesus males experience a significant
drop in play frequency (Figure 5). It could be argued therefore, that the post-pubertal increase in testosterone is related to the decline in male play. A simple hypothesis might
view testosterone as having a direct negative effect on male
play. This hypothesis would seem to adequately account for
the following ontogenetic sequence in intact rhesus males: low
testosterone levels and frequent play pre-pubertally, followed
by increased testosterone and decreased play post-pubertally.
Following the hypothesis of a direct negative effect of testosterone on male play, one would predict a seasonal increase
in play by intact males during the low-testosterone non-mating
season. Directly contradictory evidence on this point suggests that the direct-effect hypothesis is inadequate. Gordon
et al. (1976) reported on the sexual, agonistic and play behavior of adult rhesus males living in a heterosexual social
group at the Yerkes Regional Primate Research Center. These
males showed a clear seasonal cycle in testosterone production, and a significant negative correlation between the frequency of play and testosterone concentration. In contrast,
the control males of the present study failed to show a significant increase in play during their low-testosterone non-
mating seasons (see Section III, C.3).

An alternate and more complex hypothesis is that the postnatal production of adult levels of testosterone is a key factor in the attainment of both physical and behavioral maturity
by males, and that as one part of behavioral maturity, male
play declines to extremely low frequencies.

We have data on the growth of our control and castrated
males beginning when the monkeys were about 36 months of age.
Figure 12 presents data on the sitting height (crown - rump
length) of both male populations. These and other bodily
measurements (unpublished) were taken each time the animals
were caught for bleeding. It can be seen that the control
males experienced greater growth during their fourth and fifth
years than did the castrates. During the sixth year of life,
the control males were significantly larger than the castrated
males (60-65 months, $t = 3.10$, $p < .02$; 66-71 months, $t = 2.65$,
$p < .05$), with the differences again becoming non-significant
after age 6. The castrations thus produced a retardation in
the completion of physical growth in the experimental males,
probably because of the important role of testosterone in the
adolescent growth spurt of male primates (Blizzard *et al.*,
1974).

The play data suggest that the castrations may also have
produced retarded behavioral (psychological?) development and

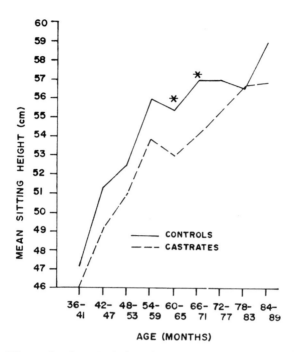

FIGURE 12. Sitting height (crown - rump length) for control and castrate males. Asterisk indicates significant inter-group difference for age period shown.

perhaps even precluded normal behavioral maturity in the experimental males. Once again, the key to these changes may have been testosterone. The importance of prenatal exposure to testosterone for normal male psychosexual differentiation has been described by Phoenix (1974b) and Phoenix, Goy and Resko (1968). Presumably, all of the males in the present study had experienced normal prenatal psychosexual differentiation in a male direction. The control males matured normally, experienced testicular descent at puberty and then, under the seasonal influence of stimulation by adult levels of the gonadal hormones (especially testosterone), they began to show a variety of adult behavior patterns, including heterosexual copulations and intra-sexual competition for estrous females. At this time, control male play was beginning a steep decline that would bring it to a rate of 0.19 episodes/male/hour by the time the monkeys were 5-1/2 - 6-1/2 years of age. It seems possible that post-pubertal, intra-sexual competition had the effect of weakening the friendly bonds between the control males and, thus, reduced the probability of male-male play.

In contrast, the castrated males never experienced puberty or stimulation by adult levels of testosterone. They did not display adult copulations and they did not compete with one another for female sexual partners. Even though the castrated males were very aggressive monkeys and were arranged in a clear dominance hierarchy, the possibility remains that male-male relationships among the castrates may have been significantly friendlier than those among the control males.

Although the play of the castrated males did show a significant age-related decline, at 5-1/2 - 6-1/2 years of age these males were playing at the rate of 1.55 episodes/male/hour (over 8 times greater than the control male rate) and almost all of their play was with other males. Our chronologically adult castrated males were playing at rates characteristic of intact males two full years younger (i.e., older juveniles and sub-adults). Their behavioral immaturity, clearly shown in regard to play, may have been related to a failure to develop normal relationships with other adult males; that failure, in turn, may have been related to their lack of postnatal testosterone stimulation.

Our studies suggest that postnatal stimulation by the gonadal hormones (especially testosterone) may be an important factor in the attainment of behavioral and psychological maturity in male monkeys. Lack of hormonal stimulation may result in atypical behavioral frequencies and atypical adult relationships.

ACKNOWLEDGMENTS

We thank the staff of the Caribbean Primate Research Center, and especially Mr. John Herbert and the crew at the La Parguera colony, for valuable assistance throughout the project. We also thank the following persons for assistance with the operations or with data analysis: Drs. William Kerber, John Vandenbergh and Jerry Hulka, and Ms. Pauline Valcourt. Dr. Jerry Robinson of the Wisconsin Regional Primate Research Center conducted all hormone assays, and we are very grateful for his contribution to the project.

REFERENCES

Barash, D. P. "Sociobiology and Behavior". Elsevier, New
    York (1977).
Blizzard, R. M., Thompson, R. G., Baghdassarian, A., Kowarski,
    A., Migeon, C. J., and Rodriguez, A. The interrelation-

ship of steroids, growth hormone, and other hormones on
pubertal growth, *in* "Control of the Onset of Puberty"
(M. M. Grambach, G. D. Grave, and F. E. Mayer, eds.),
pp. 342-366. John Wiley and Sons, New York (1974).

Cochran, C. A., and Perachio, A. A. Dihydrotestosterone propionate effects on dominance and sexual behaviors in gonadectomized male and female rhesus monkeys. *Horm. Behav.*
8, 175-187 (1977).

Conaway, C. H., and Koford, C. B. Estrous cycles and mating
behavior in a free-ranging band of rhesus monkeys. *J.*
Mammal. 45, 577-588 (1965).

Conaway, C. H., and Sade, D. S. The seasonal spermatogenic
cycle in free ranging rhesus monkeys. *Folia primat. 3*,
1-12 (1965).

Dolhinow, P. J., and Bishop, N. The development of motor
skills and social relationships among primates through
play, *in* "Minnesota Symposia on Child Psychology, Vol. IV"
(J. P. Hill, ed.), pp. 141-198. University of Minnesota
Press, Minneapolis (1970).

Duvall, S. W., Bernstein, I. S., and Gordon, T. P. Paternity
and status in a rhesus monkey group. *J. Reprod. Fertil.*
47, 25-31 (1976).

Gordon, T. P., Rose, R. M., and Bernstein, I. S. Seasonal
rhythm in plasma testosterone levels in the rhesus monkey
(*Macaca mulatta*): a three year study. *Horm. Behav. 7*,
229-243 (1976).

Hall, K. R. L., and DeVore, I. Baboon social behavior, *in*
"Primate Behavior: Field Studies of Monkeys and Apes" (I.
DeVore, ed.), pp. 53-110. Holt, Rinehart and Winston, New
York (1965).

Hausfater, G. Dominance and reproduction in baboons (*Papio
cynocephalus*). A quantitative analysis. *Contrib. Primatol. 7:* 1-148 (1975).

Hinde, R. A., and Spencer-Booth, Y. The behaviour of socially
living rhesus monkeys in their first two and a half years.
Anim. Behav. 15, 169-196 (1967).

Hotchkiss, J., Atkinson, L. E., and Knobil, E. Time course of
serum estrogen and luteinizing hormone (LH) concentrations
during the menstrual cycle of the rhesus monkey. *Endocrinology 89*, 177-183 (1971).

Koford, C. B. Rank of mothers and sons in bands of rhesus
monkeys. *Science 141*, 356-357 (1963).

Loizos, C. Play behaviour in higher primates: a review, *in*
"Primate Ethology" (D. Morris, ed.), pp. 176-218. Aldine,
Chicago (1967).

Loy, J. Estrous behavior of free-ranging rhesus monkeys.
Primates 12, 1-31 (1971).

———— The descent of dominance in *Macaca*: insights into
the structure of human societies, *in* "Socioecology and

Psychology of Primates" (R. H. Tuttle, ed.), pp. 153-180. Mouton, The Hague (1975).

Loy, J., and Loy, K. Behavior of an all-juvenile group of rhesus monkeys. *Amer. J. phys. Anthrop. 40,* 83-96 (1974).

Michael, R. P., Herbert, J., and Welegalla, J. Ovarian hormones and grooming behaviour in the rhesus monkey (*Macaca mulatta*) under laboratory conditions. *J. Endocrinol. 36,* 263-279 (1966).

Michael, R. P., and Wilson, M. Effects of castration and hormone replacement in fully adult male rhesus monkeys (*Macaca mulatta*). *Endocrinology 95,* 150-159 (1974).

Niswender, G. D., and Spies, H. G. Serum levels of luteinizing hormone, follicle-stimulating hormone and progesterone throughout the menstrual cycle of rhesus monkeys. *J. clin. Endocrinol. Metab. 37,* 326-328 (1973).

Phoenix, C. H. The role of androgens in the sexual behavior of adult male rhesus monkeys, *in* "Reproductive Behavior" (W. Montagna, ed.), pp. 249-258. Plenum, New York (1974a).

―――― Prenatal testosterone in the nonhuman primate and its consequences for behavior, *in* "Sex Differences in Behavior" (R. C. Friedman, R. M. Richart, and R. L. Vande Wiele, eds.), pp. 19-32. John Wiley and Sons, New York (1974b).

Phoenix, C. H., Goy, R. W., and Resko, J. A. Psychosexual differentiation as a function of androgenic stimulation, *in* "Reproduction and Sexual Behavior" (M. Diamond, ed.), pp. 33-49. Indiana University Press, Bloomington (1968).

Resko, J. A. Plasma androgen levels of the rhesus monkey: effects of age and season. *Endocrinology 81,* 1203-1212 (1967).

Resko, J. A., and Phoenix C. H. Sexual behavior and testosterone concentrations in the plasma of the rhesus monkey before and after castration. *Endocrinology 91,* 499-503 (1972).

Rose, R. M., Bernstein, I. S., Gordon, T. P., and Catlin, S. F. Androgens and aggression: a review and recent findings in primates, *in* "Primate Aggression, Territoriality, and Xenophobia: A Comparative Perspective" (R. L. Holloway, ed.), pp. 275-304. Academic Press, New York (1974).

Sade, D. S. Seasonal cycle in size of testes of free-ranging *Macaca mulatta. Folia primat. 2,* 171-180 (1964).

―――― Determinants of dominance in a group of free-ranging rhesus monkeys, *in* "Social Communication Among Primates" (S. A. Altmann, ed.), pp. 99-114. University of Chicago Press, Chicago (1967).

SOCIAL PLAY IN RHESUS MACAQUES (Macaca mulatta):
A CLUSTER ANALYSIS[1]

Euclid O. Smith

Yerkes Regional Primate Research Center and
Department of Sociology and Anthropology
Emory University
Atlanta, Georgia

Martin D. Fraser

Department of Mathematics
Georgia State University
Atlanta, Georgia

I. INTRODUCTION

Problems frequently arise in the study of social behavior
when one is called on to operationally define particular categories of behavior. Clearly, play is one such category. Gilmore (1966) notes that play seems to represent a definition-
ally impossible "wastebasket" category of behavior. It is
quite clear that play is an easily recognizable behavioral
class (Poole and Fish, 1975), as evidenced by the high interobserver agreement when animals are playing (Anderson and
Mason, 1974; Loizos, 1966, 1967; Miller, 1973; Sade, 1966;
Symons, 1973). However, there are real questions as to the
specific behavioral characteristics which comprise play.
The identification of specific behavioral elements within
play is a difficult task, particularly among primates, because

[1]This research was supported in part by Grant RR00165,
United States Public Health Service; Grant MH20483, National
Institute of Mental Health; and a Dissertation Fellowship from
the Graduate School, Ohio State University.

79

of the rapidity and temporal lability of the behavior. It is
precisely the temporal characteristics of the behavioral elements which comprise play interactions which have caused considerable controversy in the literature. For example, Loizos
(1967:179), among others, notes that "...play has no formal-
ized sequence of events, such that action A will always be
followed by action B, C, or D. In play, depending upon the
feedback from the object or the social partner, A may be followed with equal likelihood by B or Z." On the contrary,
Müller-Schwarze (1971) found sequential stability and predictability in play sequences of blacktail deer.

Furthermore, if we review the literature we find a wide
variety of definitions of play. Altman (1966) noted that play
had never been satisfactorily defined and, in fact, was the
most neglected dimension of animal social behavior. Given the
lack of agreement over what precisely constitutes a definition
of play, a structuralist position (e.g., Fagen, 1974) has been
adopted which addresses not the underlying behavioral mechanisms or adaptive significance of play, but considers the form
of the behavioral elements which comprise play.

In any consideration of the structure of play, one must
come to grips with the problem of relying on certain intuitive
typologies to define this complex behavioral element. It is
tempting to rely exclusively on certain characteristics of
play which allow us to achieve high interobserver reliability
measures as to when animals are playing. However, researchers
should strive to more objectively define such an event.

This paper attempts to empirically define social play in
rhesus macaques by analyzing the social behavioral repertoire
of young animals using a multivariate statistical procedure to
arrive at a definition of social play in terms of the motor
patterns exhibited.

II. METHODS AND MATERIALS

A. *Study Site*

This study was conducted at the Yerkes Regional Primate
Research Center Field Station near Lawrenceville, Georgia, and
involved behavioral observations of two groups of rhesus macaques, *Macaca mulatta*. These two groups (designated R-9 and
R-12) are long—standing, intact social groups and were selected for the study because of large size and age/sex composition (R-9) and comparative value of observations on the same
species (R-12). Long-term genealogical records are available
for both groups (see Tables I and II), which allowed careful
control of selection of subjects for observation. Although

TABLE I. *Genealogical Relations and Age/Sex Composition of R-9 Group*

Adult	1970	1971	1972	1973	1974	1975
NB♀		ZF♂	QG♀[a]	QH♂[a]	DJ♂[a]	GK♀
RA♀	GF♀	EG♀		TH♂[a]	CJ♂[a]	ZK♀
	GF♀			DI♀[a]		RK♀
		EG♀				UK♀
BB♀	IF♀			YH♂		PK♀
	IF♀				LI♀[a]	SK♂
QA♀	LF♀		KG♂[a]			
	LF♀				MI♀	MK♂
VA♀	PF♀		SG♂[a]	BI♂	EJ♀[a]	NK♂
	PF♀				JI♀	KK♂
UB♀			NG♂[a]	WH♀[a]	GJ♀[a]	JK♂
QD♀		VF♂		LH♀		
SA♀	QF♂				QI♀	EL♀
WC♀	JF♀			OH♀[a]	UI♂[a]	BL♂
	JF♀					VJ♂
OC♀	OF♀		AH♀[a]	NH♂	TI♂[a]	WJ♀[b]
	OF♀				ZI♂	
IE♀			UG♀		II♂	DL♂
YC♀	NF♀		HG♀[a]			
	NF♀					YJ♂
			HG♀			UJ♀[b]

Continued on following page

Adult	1970	1971	1972	1973	1974	1975
RC♀	———————————————		BG♂ ———————————————			TJ♂
BD♀	————————	TF♂ ———————		PH♂ ————————		HK♂[b]
JB♀	————————	WF♀ ———————		KH♂[a]———	HJ♀[a]———	WK♀
		WF♀ ———————————			VI♀ ———	OK♂
VC♀	——— RF♀ ————————			GI♂ ————————		ZJ♂
	RF♀ ————————				SI♀ ———	YK♀
FB♀	———————————————			CI♀[a]———		TK♂
TD♀	———————————————			FI♂[a]		
	HF♀ ————————				NI♂ ———	QK♀
	MF♀ ————————				WI♀[b]———	LK♂[b]
	KF♂ —————————	PG♀[a]				
		UF♂ ——— YG♂				
WA♀						
PC♀						
VD♀						
WD♀						
UA♀						
PD♀						
UE♀						
PA♀						
WB♀						
RE♀						
SE♀						

Continued on following page

Adult	1970	1971	1972	1973	1974	1975	
PE♂							
QE♂							
CE♂							
OB♂							
QB♂							
7♂	2♂	4♂	5♂	9♂	7♂	16♂	50
27♀	10♀	2♀	5♀	5♀	10♀	10♀	69

[a] Subjects.

[b] Died during course of study.

the subjects were being used concurrently in other experiments, manipulations had no observable effect on their social interactions. No changes in group composition were noted during the study except through natural causes.

The animals are housed in identical 38.5 m x 38.5 m open-air enclosures surrounded on one side by a 4.85 m high sheet metal wall and on three sides by 1.75 m of 5 cm chain link fence surmounted by 3.1 m of sheet metal [see Gordon and Bernstein (1973) for diagram of compound]. The open-air enclosures are attached to 9.3 m x 3.1 m x 2.2 m indoor quarters. Two animal-operated swing doors allow access between the indoor quarters and the outside compound. There are indoor and outdoor observation posts from which all sections of the compound can be seen (see Figure 1).

Furthermore, genetic variability between these two groups was minimized, as the smaller group (R-12) was formed from the larger group (R-9). The R-9 group was formed in 1969, with the first births occurring in 1970. The R-12 group was split from the R-9 group in 1970, with the first births in 1971 [see Bernstein, Gordon and Rose (1974a,b) for a full discussion of these group formations].

Dolhinow and Bishop (1970) have noted that play is one of the most difficult behavior patterns to study because of rapidity and variability of expression. Any detailed study of social play is destined for failure unless observation conditions are optimal. This is clearly not the case in most field

TABLE II.   *Genealogical Relations and Age/Sex Composition of R-12 Group*

Adult	1971	1972	1973	1974	1975
AE♀	YF♀	RG♀ᵃ			
	YF♀				BK♀
OA♀		MG♀ᵃ	RH♂ᵃ	IJ♂ᵃ	AL♀
OD♀			IH♀ᵃ	RI♀ᵃ	XK♂ᵈ
SB♀		LG♂ᵃ,ᵇ	JH♂ᵃ	FJ♀ᵃ	IK♂
SC♀			AI♂ᵃ	BJ♂ᵃ	CK♂
SD♀		EH♀ᵃ		YI♂ᵃ	
TC♀			EI♀ᵃ		VK♂
TE♀		FG♂ᵃ	SH♀ᵃ	PI♂ᵃ	EK♂
VE♀	SF♀	OG♂ᵃ,ᶜ	MH♀ᵃ		
	SF♀				AK♀ᵈ
YD♀			ZH♂ᵃ		FK♀
JC♂					
BC♂					
2♂	0♂	3♂	4♂	4♂	5♂	18
10♀	2♀	3♀	4♀	2♀	4♀	25

ᵃSubjects.

ᵇAdopted mother MA (removed prior to study).

ᶜAdopted mother SC.

ᵈDied during course of study.

FIGURE 1. View of the outdoor animal enclosure from the photographic platform at the Yerkes Regional Primate Research Center Field Facility.

studies. However, to appreciate the variability of this highly social behavior, a study of play must involve observations of individuals living within a social group. Indeed, a compromise on either of these key issues would seem to limit results at the outset. Therefore, the use of captive groups of primates, housed in enclosures which allow spatial mobility, seems to offer the most suitable opportunity for the study of social play. It is precisely these factors which influenced the choice of the Yerkes Regional Primate Research Center Field Station as a study site. Further, the accessibility of two intact groups of rhesus macaques, as well as the quantity of existing data from field, laboratory and provisioned colonies, on social behavior in *M. mulatta*, argued strongly for their choice as study species.

B. *Study Species*

Many researchers have noted that play occurs most frequently in young animals (Bekoff, 1972; Poirier, 1972; Poirier and Smith, 1974; Smith, 1972). Symons (1973) and Hinde and Spencer-Booth (1967) noted specifically that, during the first

2-1/2 - 3 years of life for rhesus macaques, maximal social
play is witnessed. For that reason, a sample of young individuals was selected from each group for intensive study. Subjects were selected using the following criteria, wherever
possible: 1) consideration of maternal rank, determined from
existing data, individuals were selected from both high- and
low-ranking mothers; 2) males and females were selected to assess the importance of gender differences on the expression of
play; 3) as often as possible, siblings were selected from
various matrilines to investigate certain dimensions of peer
vs. sibling preference for play partners; 4) individuals were
also selected which did not have a mother in the group to assess this social variable on expression of play; and 5) only
individuals in their first, second and third year of life were
selected. It was possible to select subjects in the large
group (R-9) (n = 119) with all these criteria in mind; however, the size (n = 43) of the smaller group (R-12) precluded
consideration of some of these criteria. In fact, to achieve
a sufficient sample size, all individuals in their first, second and third year of life in the R-12 group were included as
subjects (see Tables I and II for subject and sex designation).

C. *Initial Pre-test Observations*

    Initially, three and one-half months (approximately 350
hours) of observation were required to accurately identify
each individual in the large group (R-9). Later in the study,
an additional 40 pre-test observation hours were required to
individually identify each member of the small group (R-12).
Although all animals were tattooed with individual alphabetic
codes, these were so small as to be of little use in a large
open-air enclosure. Individuals had to be recognized by discrete morphological characteristics and/or individual behavior
patterns.
    Although a time-consuming process, this procedure allows
collection of data on individual animals in a manner clearly
superior to simple identification of age/sex classes. To aid
in the identification process, some of the animals were captured and given a unique shave code.

D. *Behavioral Inventory*

    Any study of social behavior must begin with a well docu-
mented inventory of behavioral responses; the study of social
play is no exception. "The basis for ethological investigation is the ethogram, the precise catalogue of all the behavior patterns of an animal" (Eibl-Eibesfeldt, 1975:11). In

practice, functional units are chosen which are neither too
large nor too small, but consistent in form. However, the description of a behavior pattern is never complete in actuality,
for the observer makes certain judgmental decisions on the relative importance of behaviors (Eibl-Eibesfeldt, 1975).

Wiepkema (1961) selected behavioral units utilizing several
different criteria: 1) easily measurable, 2) not too rare in
occurrence, 3) biologically meaningful, and 4) not entirely
correlated with other variables. Norton (1968) noted that a
behavioral unit consists of one or several movements occurring
simultaneously or in immediate sequence with a high predictability. According to Kummer (1957), a behavioral unit is an
essential core movement (Krenbewegung) which can be accompanied
by typical or accessory movements. Hopf (1972) employs a similar definition in her study of squirrel monkeys. These units
should be defined operationally, although the 'gestalt' of them
cannot, and should not, be completely disregarded (Lorenz,
1959). The names of the units should be thought of as labels,
not as a *priori* interpretations (Hopf, 1972).

van Hooff (1973) has noted that the question is how far to
split movements and postures, and where to lump movements and
postures to create a catalogue of meaningful units. Most catalogues, especially in laboratory studies, are relatively small,
so that a great deal of lumping and/or selection inevitably
must have occurred (Plutchik, 1964). "Most students of primate
behaviour have not worried too much about this and have presented catalogues in a matter-of-fact manner" (van Hooff, 1973:
81).

According to Altmann (1965:492), "If one's goal is to draw
up an exclusive and exhaustive classification of the animal's
repertoire of socially significant behavior patterns, then
these units are not arbitrarily chosen. To the contrary, they
can be empirically determined. One divides up the continuum
of action wherever the animals do. If the resulting recombina-
tion units are themselves communicative, that is, if they affect the behavior of other members of the social group, then
they are social messages. Thus, the splitting and lumping that
one does is, ideally, a reflection of the splitting and lumping
that the animals do. In this sense, then, there are natural
units of behavior." van Hooff (1973) continues that there are
a number of behavioral elements which have characteristics
which differentiate them from other elements, and cannot be
divided into independently occurring subdivisions.

Social behavior of primates, then, to some extent, forms a
graded continuum (e.g., touch, push, pull, hit, slap) and may
exhibit similar motor elements which vary in intensity. Also,
there may be a phylogenetic relationship in the degree or sub-
tlety of the gradations, with pongids exhibiting the most in-
tergraded behavior (Marler, 1959, 1965). As von Cranach and

Frenz (1969) note, the use of too broadly defined, as well as
too narrowly defined, behavioral units can reduce the validity
of results. Caution should be exercised to identify socially
and biologically meaningful elements.

Prior studies of rhesus macaques have included catalogues
of social behavior in varying stages of completeness (Altmann,
1962; Carpenter, 1942; Chance, 1956; Hinde and Rowell, 1962;
Hines, 1942; Rowell and Hinde, 1962). However, as Reynolds
(1975) points out, comparisons of the results are difficult
due to a variety of factors (i.e., different aims, lumping and
splitting differences, usage differences). For example, Altmann (1962) employs one category, "plays with", and notes,
"...play is not a single behavior pattern, but rather a complex form of behavior interactions involving many other behavior patterns in highly modified form" (Altmann, 1962:376-377).
Chance (1956:4) notes that play included "(1) grasping of one
animal by another, (2) attempts to bite, (3) pulling, in attempts to displace or turn over, (4) rolling over, (5) jumping: (a) in the air by itself, (b) and catching at tail or
grasping, (c) and looking through its legs, (d) and losing its
balance (repeatedly), (6) sliding off tree trunks, etc., (7)
imitative, incomplete activities: (a) pelvic movements, (b)
threats, (c) equilibratory movements and gestures, (d) biting." Also, as a distinct category, Chance (1956) included
bathing activities of juveniles which, by many standards,
would be playful (i.e., Symons, 1973). While Hines (1942)
recognizes two forms of play, movement play and social play,
he does not present an adequate operational definition for either type.

From these descriptions, considerable variability in the
development of an ethogram for rhesus macaques can be noted.
From these published data, a behavioral inventory already in
use at the Yerkes Regional Primate Research Center for observations on rhesus macaques (Bernstein, personal communication)
and some 10 hours of filmed interactions, a catalogue of 118
behaviors was developed [see Smith (1977) for complete listing
and operational definitions]. Included in this inventory are
a variety of behaviors, not all of which are playful. In
fact, there are no molar (large, inclusive/functional) behavioral categories in the inventory, for it was felt that pre-
mature lumping (e.g., sexual, agonistic, play) would obscure
the variability in expression of patterns of interest.

Frequently, behavior patterns have been classified in
terms of their function; however, this presupposes a knowledge
of causation which is one of the aims in the study of behavior
(van Hooff, 1973). Ethologists have cautioned against the use
of functional classifications on an *a priori* basis precisely
for this reason, not only in classification, but also in the
description of the behavioral elements (Baerends, 1956; Hinde,

1966; Tinbergen, 1963; van Hooff, 1973). Furthermore, since formidable definitional problems exist with social play (cf., Rowell, 1966, 1967, 1969) as a molar behavioral category, every attempt was made to define as precisely as possible the component behaviors within a play bout.

E. *Data Collection Procedures*

Observations began 16 October, 1974, and terminated 17 October, 1975. During this period, over 750 hours of observation yielded 160 hours of quantified data (120 hours for R-9; 40 hours for R-12). Data were collected only when the ambient temperature was below 29.4° C and above 7.2° C, for it has been demonstrated (Bernstein, 1972, 1975; Bernstein and Mason, 1963) that temperatures outside this range markedly affect general activity levels and, thereby, the expression of social play. Play does occur at temperatures outside this range, but these limits were observed in order to maximize data, by sampling during periods of maximum activity for the entire group. Data were also not collected in rainy weather, as this condition has been demonstrated to minimize social play (Bernstein, 1972, 1975; Oakley and Reynolds, 1976). All observations were made from the outdoor photographic platform, with animal access to indoor quarters restricted. Therefore, all subjects were constantly in view.

F. *Sampling Techniques*

The focal animal sampling technique (Altmann, 1974) was used during the entire study, although qualitative notes were taken to capture aspects of interactions not usually scored. The focal animal technique is particularly appropriate for the study of social play, as it is designed to record all occurrences of specified interactions of an individual during each sampling period. Furthermore, under some conditions, one may assume that a complete record is obtained of the focal animal's actions as well as the actions directed toward him by others.

Often results of focal animal sampling, and other techniques for that matter, are used to make statements about frequency of interactions when they are actually statements about rates (Altmann, 1974). Recognizing that there are important theoretical differences concerning the rate of responses and duration, data were collected on the time of onset of each interaction and the termination of those durational responses. For example, slaps, bites, jump ons, etc., are of relatively short duration and are significant in terms of their frequency

of occurrence, while grooms, maternal behaviors, etc., are important for their duration as well. Consequently, data were
recorded on: 1) the focal animal, 2) the interactant, 3) the
sequence of interactions, 4) the time of occurrence, 5) direction of interaction, and 6) the onset and termination of durational behaviors.

The 44 focal animals were observed individually for six-
minute observation periods. The duration of the testing interval was, as are most conditions in design of a study, a
compromise between prolonged sessions in which the behaviors
of an individual could be monitored in half-hour blocks, for
example, and short sessions but of much greater frequency
(e.g., 30 seconds). Since the sequence of behaviors during
the testing sessions was of interest, the duration of the test
interval should be of sufficient length to include an adequate
sample of the longest sequences of interest (Altmann, 1974).
Independent measurement of the average duration of 100 play
bouts [defined by criteria suggested by Bernstein (1975)]
yielded a mean of slightly less than 1-1/2 minutes. Therefore, it was concluded that a six-minute test session would be
of sufficient duration for adequate sampling. Individuals
were chosen at random from the subject list without replacement for a given day's observations, and observed for a six-
minute test session. If, however, an interesting interaction
sequence was in progress at the termination of the test interval, data were collected until the termination of the interaction, but used only to add qualitative detail, and are not included in any statistical analysis.

In sum, focal animal observations yielded a total of 1,600
six-minute observation sessions (1,200 for R-9; 400 for R-12).
In other words, during the course of the study, each focal
animal in the R-9 group was observed for five hours, while
each subject in the R-12 group was observed for two hours.
Observations were conducted when the ambient temperature was
within the predefined range; however, time of day and number
of observations per month were not controlled, although efforts were made to sample behavior throughout the diurnal and
annual cycle.

Because of the rapidity and temporal lability of much of
rhesus behavior, the traditional checklist approach was not
employed. However, the same cautions were used in tape recording the data as would be employed in the use of the checklist approach [vide Hinde (1973) for details]. Interactions
were dictated onto cassette tapes using various recorders.
Observations were aided by a pair of 7 x 50 wide angle binocu-
lars.

Tape recorded data were then transcribed into a sequential
event format of, basically, who, does what, to whom. Using
the original data tape and the typed scenario, the times for

the onset of all behaviors and the termination of some (e.g.,
maternal, groom, huddle) were recorded by playing the tape in
real time and noting times after the start of each session.
Therefore, not only was the time recorded in relation to the
onset of the test session, but also in relation to a 24-hour
clock.

During the course of the study, periodic filming allowed
continual reevaluation of categories being scored, as well as
preventing 'observer drift' as greater facility was acquired
in making observations (Poole, 1973). Identical testing and
data collection procedures were employed for both groups.

Data were then coded onto Fortran coding forms and then
keypunched. Observations over the twelve-month period on both
groups yielded in excess of 40,000 separate interactions.
After keypunching, data were subjected to a variety of editing
programs to detect coding and/or keypunching errors.

G. Multivariate Techniques for Analysis of Behavioral Data

Recognition of problems of classification of behavior and
search for underlying causal mechanisms can be attributed to
the work of classical ethologists on predominantly non-mammal-
ian forms (vide Hinde, 1966; Marler and Hamilton, 1966; Tinbergen, 1942). These early studies are significant, for they
attempt to present an analysis of the structure of the behavior on an empirical rather than intuitive basis.

A structural analysis of behavioral data and the classification of behavioral elements on the basis of temporal relationships seems the logical preliminary to the investigation
of the causal factors and functional aspects of behavior (van
Hooff, 1973). For such analysis, many researchers have turned
to multivariate statistical techniques, largely derived from
information theory. While Haldane and Spurway (1954) were the
first to use information theory on nonhuman behavioral data, a
number of researchers have followed, employing a variety of
techniques [cf., Baerends and van der Cingel (1962) for displays of common herons; Dingle (1972) for aggressive behavior
in stomatopods; Fentress (1972) for grooming in mice; Slater
and Ollason (1972) in a study of behavior of male zebra finches; Steinberg and Conant (1974) for intermale behavior of
grasshoppers; and Wiepkema (1961) in a study of reproductive
behavior in bitterlings].

Several researchers have used multivariate analysis of the
temporal relationship of behavior in primates (Altmann, 1965,
1968; Chamove, Eysenck and Harlow, 1972; Jensen, Bobbitt and
Gordon, 1969; Locke et al., 1964; Maurus and Pruscha, 1973;
Morgan et al., 1976; Pruscha and Maurus, 1973; van Hooff,
1970, 1973). However, the possibilities for use of these

techniques in understanding and explaining behavior seem in
their infancy.

Basically, it seems that there exist four alternative approaches to the problem of the sequential and/or temporal
analysis of behavioral data:

   1. *Correlation Techniques.* There exist a variety of
techniques for measuring frequencies of different behavior
patterns within a series of time units to determine whether
they are positively or negatively correlated (e.g., Andersson,
1974; Baerends, 1956; Bekoff, n.d.; Delius, 1969; Heiligenberg, 1973). Positive correlations indicate relatively similar causal bases, low correlations indicate the causal bases
are unrelated, and negative correlations indicate an inhibiting relationship (Hinde, 1966). However, several problems may
plague the researcher: (a) if the number of behavioral elements is large, the interpretation of a great number of correlation coefficients that can be obtained by this method can be
difficult (van Hooff, 1973); (b) if the probability of the occurrence of a particular set of behaviors does not remain constant over time, nonstationarity becomes a problem, but not as
acute as in other techniques (for further discussion of stationarity, see factor analysis technique). A check of the
data for stationarity is a first requirement (Slater, 1973),
although diurnal cycles (e.g., Aschoff, 1967; Palmgren, 1950)
and short-term cycles (Richter, 1927; Wells, 1950) may render
the stationarity assumption untenable; (c) the main difficulty
is that results depend on the choice of time unit (Slater,
1973). Although two acts are significantly correlated at one-
half hour intervals, this is not necessarily true when data
are organized in 10-second intervals.

   2. *Markov Analysis.* A sequence of behaviors can be described by a first-order Markov chain if the probabilities of
different acts depend on the immediately preceding act and not
on any earlier ones (Billingsley, 1961; Cane, 1961). However,
Markov chains can extend longer than just the preceding interaction. An *nth* order approximation is possible, for a species
with a repertoire of r mutually exclusive patterns of social
behavior will have r^{n-1} states, corresponding to the *n-1* event
before the *nth* event (Altmann, 1965). This type of analysis
has been used to describe social communication in rhesus macaques (Altmann, 1965), courtship patterns in glandulocaudine
fish (Nelson, 1964), sequences of bird song in the cardinal
and the wood pewee (Chatfield and Lemon, 1970), and suggested
as a technique for analyzing social play (Bekoff, 1975).
Here, too, methodological problems confront the researcher:
(a) as noted before, stationarity poses considerable problems.
As Slater (1973:145) noted, "...assuming stationarity of

behavioral events is tantamount to ignoring the possibility that motivational events occur." The effect of diurnal, circadian or seasonal cycles may not allow the assumption of stationarity; (b) Bekoff (n.d.) notes that conditions of observation of social behavior often preclude the use of Markov analysis because of the stationarity problem [e.g., two individuals' interaction, data from different individuals is lumped together (Chatfield, 1973)].

   3. *Factor Analysis*. Factor analysis is a multivariate statistical method, not primarily concerned with sequencing, but used to detect association patterns between behaviors. Factor analysis has been used by a number of researchers (Baerends and van der Cingel, 1962; Baerends *et al.*, 1970; Chamove, Eysenck and Harlow, 1972; Locke *et al.*, 1964; van Hooff, 1970; Wiepkema, 1961). Using this method, a small number of hypothetical variables are extracted, the existence of which could account for most of the observed correlations between acts. It is assumed that the behavior patterns (variables) do not depend causally on each other, but only on the underlying postulated factors (Blalock, 1968; Slater, 1973). The choice between factor analysis and Markov analysis appears to depend mainly on whether the researcher believes "sequence effects" or "motivational state" to be most important in determining the relationship between behaviors. If "sequence effects" are felt to be more important, then it is useful to describe the probability of the occurrence of a particular act at a point in time, in terms of the sequence of acts which preceded it. If, on the other hand, the emphasis is on motivational state, an association between acts is taken to mean that there are common causal factors underlying the behaviors, and the factor on which they have a high loading may represent a motivational state [e.g., aggressive, sexual, nonreproductive (Wiepkema, 1961); affinitive, play, aggressive (van Hooff, 1970)] (Slater, 1973).
   Several criticisms of the use of factor analysis in ethological research have been discussed by Andrew (1972), Overall (1964) and Slater and Ollason (1972). Basically, the question is the interpretation of the underlying variables. Wiepkema (1961) referred to factors as "tendencies", Baerends *et al.* (1970) note that factors were areas of high density within the causal network, while van Hooff (1970) called his factors main motivational systems. Slater (1973:145) notes that, "...it is doubtful whether the extraction of factors which are themselves of complex causation advances understanding."

   4. *Cluster Analysis*. Morgan *et al.* (1976) note that there are a number of ways of tackling multivariate data.

Cluster analysis seems preferable to other techniques, as it makes fewer assumptions about the data and is, therefore, easier to understand. Maurus and Pruscha (1973) argue strongly for the use of cluster analysis over other techniques, especially factor analysis, for the following reason. With factor analysis techniques, frequency data are interpreted as points in Euclidian space, which is against the nature of the data. van Hooff (1970) subjected behavioral data to a factor analysis in the following manner: for each cell of the transition matrix (a matrix where the behaviors of the inventory were listed down the rows and across the columns, with the frequency of two acts occurring in succession noted at the intersection). For example,

Behavior from
Inventory
(Succeeding)

		A	B	C	
	A	3	7	8	18
Behavior from Inventory (Preceding)	B	2	1	5	8
	C	8	3	2	13
		13	11	15	39 = N

which can be read as A preceded A on three occasions; A preceded B on seven occasions, etc. The expected frequency for each cell in our example could be calculated by taking the row total for a particular cell times the column total for that cell divided by the total N.

$$e = \frac{\text{row total for particular cell X column total for particular cell}}{N}$$

o = observed value for particular cell

N = matrix total

To minimize the effect of random variation, van Hooff (1973: 132) used the following:

$$q = \frac{o-e}{\sqrt{e}} \quad \text{(eccentricity coefficient)}$$

o = observed value for particular cell

Then (van Hooff, 1973), a matrix was constructed consisting of
the Spearman rank-correlation coefficients of two behavior
units which were then subjected to factor analysis, i.e.,
Spearman correlation coefficients versus the actual raw data,
may offer difficulties in interpretation of results (Maurus
and Pruscha, 1973).

van Hooff (1973) further subjects his data to a cluster
analysis using the hierarchical method suggested by McQuitty
(1966). However, for the present analyses, the technique sug-
tested by Orloci (1967, 1968) in a study of plant communities
seems more appropriate because, as Maurus and Pruscha (1973:
122) note, "...the original frequencies are preserved at every
step and the data are never subjected to an algorism which is
against the nature of these data."

H. Single Link Cluster Analysis

Following the technique described by Maurus and Pruscha
(1973), behavioral observations were reduced to a transition
matrix, n(h,k) is the number of times actions h and k occur in
succession in a series of observations (see preceding example).
Furthermore, these actions must have occurred within one 6-
minute test session for a given focal subject. The total frequency of the matrix, n, equals the total observations in the
study.

There are many alternative cluster analysis procedures
[vide Cormack (1971) for a review], but the "agglomerative
clustering method", or single link technique, was chosen.
This technique has been applied to diverse biological data
(Cole, 1969; Orloci, 1967, 1968; Sokal and Sneath, 1973). Using this procedure, each behavior forms a different class, m
(a subset of behaviors from the original catalogue), at the
starting procedure (step 0). Thus, at step 0, m = N, the number of behaviors in the catalogue equal the number of classes,
m. In the case of the present analysis, m = 118.

In the first step, two behaviors are chosen from the catalogue and combined into one class: m = N-1. This procedure
is then repeated until all actions are combined into one
class: m = 1. As a result, a hierarchy of classes becomes
more and more comprehensive, and can be depicted in a dendogram (see Figure 2).

The choice of two classes, m, for combination can be il-
lustrated in a transition matrix (M1). Given m classes of behaviors A_1, A_2,...A_m, the transition matrix can be represented
as:

$$n(A_i, A_j) i, j = 1,...,m$$

that pair out all of $\binom{m}{2}$ pairs A_i, A_j, $i \neq j$, is combined into
a new class for which the new transition matrix (the matrix
generated after the combination of A_1, A_j) shows the maximum
transinformation. In other words, behaviors are paired together (clustered) by taking all combinations of pairs within
the matrix and collapsing those two that have maximum transinformation at that step.

The transinformation, T, of a frequency matrix is a measure of the relation between the row and column categories. The
properties of T are seen at its extreme values. When T = 0,
the sequence of behaviors is stochastically independent; when
T reaches a maximum, each behavior is fully determined by the
preceding one [for a more detailed review of the properties of
T, see Khinchin (1957)]. Maurus and Pruscha (1973) use this
method, establishing a hierarchy of classifications using the
transinformation as the criteria of selection when the observation data form a contingency table. The transinformation
measure, T (Pruscha and Maurus, 1973), is the quantity

$$T = \sum \frac{n_{ij}}{n_{..}} \log \frac{n_{ij} n_{..}}{n_{i.} \ n_{.j}}$$

where frequencies n_{ij}, $n_{i.}$, etc., are defined as usual in the
case of transition matrices (Billingsley, 1961; Fano, 1961).
The application of this technique to a matrix in which the
rows and columns represent the same set of characters, the
catalogue of behaviors, requires a few modifications [vide
Maurus and Pruscha (1973:110) for details].

For purposes of analysis, only social behaviors (i.e., not
self-directed, or used to signal the termination of a behavior) occurring with more than 0.01% of the total frequency of
interactions (N = 47) were used. Interactions for all age/sex
classes were lumped together for purposes of this analysis,
although admittedly variation clearly exists. Therefore, the
original matrix (step 0) was a 47 x 47 table.

One of the problems in using the cluster analysis method
is determination of the best possible clustering arrangements,
those most closely related to the realities of behavior.
Maurus and Pruscha (1973) note difficulties here, and resort
to the use of a probabilistic interpretation of the transinformation of a transition matrix. Following Anderson and
Goodman (1957), they use a chi-square function where, with a
sufficiently large n, 2T is a chi-square with $(N-1)^2$ degrees
of freedom. However, in this study, the X^2 values fall into a
region which cannot be distinguished from 1, so an alternative
approach was adopted as a way of interpreting the best link-
ages for the clusters. As an alternative, because of the
large number of degrees of freedom (df = 2116, at step 0), the

expression $\sqrt{2\chi^2}$ - $\sqrt{2df-1}$ (Blalock, 1972) was used as a normal deviate with unit variance, and the corresponding Z statistic was calculated (see Figure 2). At the step where the Z value reached maximum, the best clustering of the data had been reached. The Z values at each fifth step in the clustering procedure are shown for clarity on the ordinate of the dendogram (Figure 2). It should be noted that the use of the Z statistic, or other techniques, is simply an indicator of the best cluster and, as such, is superior to an intuitive guess, but is not a true inferential statistical technique.

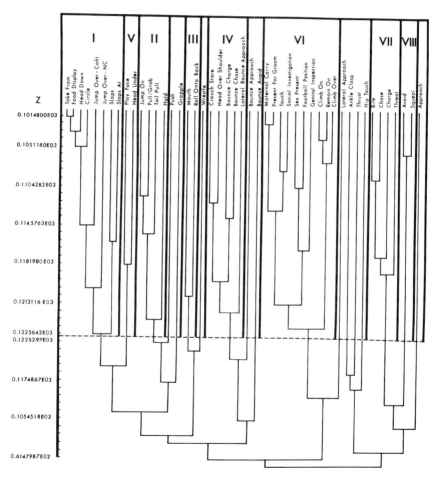

FIGURE 2. Single link dendogram of rhesus social behaviors used in the analysis. The dotted horizontal line represents the optimum clustering step. Z scores are shown for every fifth clustering step. The dotted vertical lines partition the various clusters.

III. RESULTS

    Data on over 14,000 social interactions for the R-9 group
were analyzed by the cluster analysis technique previously
described. Only social behaviors with a frequency of greater
than 0.01% of the total were selected from the catalogue (see
Table III), for it was felt that the inclusion of low frequency events would add little to the analyses. Maurus and
Pruscha (1973) in a study of social communication in captive
squirrel monkeys included only those behaviors occurring with
a frequency of greater than 0.05% in their cluster analysis,
for behaviors with lower frequency could bias the data. Also,
durational behaviors (e.g., maternal, groom), maintenance behaviors (e.g., feeding, drinking), and some clearly non-playful behaviors (e.g., huddle) were excluded from these analyses.
    The results of the cluster analysis are depicted in Fig-
ure 2, which is a graphic representation of the clusters of
the 47 selected behaviors. This graphic representation follows the general convention suggested by Maurus and Pruscha
(1973), Morgan et al. (1976) and Jardine and Sibson (1971) for
single link cluster analysis. The behaviors have been arranged in such a fashion to best represent the clustering patterns. As can be seen at the final step in the clustering
procedure, all behaviors are linked together, but this is relatively meaningless, as one would expect all social behaviors
to form an integrated, interrelated system. The relations of
the behaviors, to be sure, are dependent on the level of the
analysis, but the structure is clearly of a hierarchical nature. At one level, the specific elements of the inventory
are emphasized, while the common dimensions of groupings of
elements can be seen at another (van Hooff, 1973).
    When all observations are combined, several interesting
patterns emerge. Using the horizontal dotted line in Figure 2
as an indicator of the best clusters, it is clear that several
prominent clusters of behaviors are present, while other behaviors are not linked to any of the existing clusters. For
example, take from, food display, head down, circle, jump over
with contact, jump over without contact, slaps, and slaps at
are all joined together; however, push, grapple, etc., are not
linked at the optimum cluster step, although these "isolated"
behaviors ultimately join with clusters of behaviors.

A. Play Clusters

    Lumping elements of the behavioral inventory into clusters
is helpful in allowing the association of behaviors to become
more apparent, but one is faced with the problem of assigning

TABLE III. *Social Behaviors Selected from the Behavioral
Inventory and Used in the Cluster Analysis*

Behavior[a]	Frequency						Total	
	R-9			R-12				
Take From	$\dfrac{r^e}{r_R}\ \dfrac{7}{13}$	20	$(0.14)^b$	$\dfrac{0}{2}$	2	$(0.06)^c$	22	$(0.13)^d$
Food Display	$\dfrac{8}{9}$	17	(0.12)	$\dfrac{4}{0}$	4	(0.12)	21	(0.12)
Head Down	$\dfrac{5}{11}$	16	(0.11)	$\dfrac{1}{1}$	2	(0.06)	18	(0.10)
Circle	$\dfrac{26}{2}$	28	(0.20)	$\dfrac{12}{3}$	15	(0.49)	43	(0.24)
Jump Over Contact	$\dfrac{181}{183}$	364	(2.59)	$\dfrac{31}{22}$	53	(1.57)	417	(2.39)
Jump Over Non-Contact	$\dfrac{175}{197}$	372	(2.65)	$\dfrac{49}{29}$	78	(2.32)	450	(2.58)
Slaps	$\dfrac{94}{157}$	251	(1.79)	$\dfrac{12}{25}$	37	(1.10)	288	(1.65)
Slaps At	$\dfrac{87}{42}$	129	(0.92)	$\dfrac{24}{17}$	41	(1.22)	170	(0.97)
Play Face	$\dfrac{226}{205}$	431	(3.07)	$\dfrac{72}{47}$	119	(3.53)	550	(3.16)
Head Under	$\dfrac{81}{75}$	156	(1.11)	$\dfrac{17}{8}$	25	(0.74)	181	(1.04)
Jump On	$\dfrac{79}{48}$	127	(0.90)	$\dfrac{2}{2}$	4	(0.12)	131	(0.75)
Pull/Grab	$\dfrac{107}{81}$	188	(1.34)	$\dfrac{12}{11}$	23	(0.68)	211	(1.21)
Tail Pull	$\dfrac{54}{37}$	91	(0.65)	$\dfrac{8}{7}$	15	(0.44)	106	(0.61)
Hold	$\dfrac{321}{279}$	600	(4.27)	$\dfrac{59}{73}$	132	(3.92)	732	(4.20)
Push	$\dfrac{325}{335}$	660	(4.69)	$\dfrac{48}{46}$	94	(2.79)	754	(4.33)
Grapple	$\dfrac{564}{280}$	884	(6.29)	$\dfrac{110}{69}$	179	(5.31)	1063	(6.10)

Continued on following page

Behavior[a]	Frequency R-9			R-12			Total	
Mouth	$\frac{424}{357}$	781	(5.55)	$\frac{69}{77}$	146	(4.33)	927	(5.32)
Roll Onto Back	$\frac{6}{3}$	9	(0.06)	$\frac{1}{0}$	1	(0.03)	10	(0.06)
Wrestle	$\frac{266}{117}$	383	(2.72)	$\frac{24}{13}$	37	(1.10)	420	(2.41)
Crouch Stare	$\frac{31}{17}$	48	(0.34)	$\frac{12}{1}$	13	(0.39)	61	(0.35)
Head Over Shoulder	$\frac{46}{33}$	79	(0.56)	$\frac{13}{5}$	18	(0.53)	97	(0.56)
Bounce Charge	$\frac{25}{19}$	44	(0.31)	$\frac{8}{8}$	16	(0.48)	60	(0.34)
Bounce Chase	$\frac{272}{229}$	501	(3.56)	$\frac{63}{54}$	117	(3.47)	618	(3.55)
Lateral Bounce Approach	$\frac{103}{83}$	186	(1.32)	$\frac{27}{24}$	51	(1.51)	237	(1.36)
Bounch Approach	$\frac{315}{226}$	541	(3.85)	$\frac{9}{7}$	16	(0.48)	557	(3.20)
Bounce Avoid	$\frac{287}{236}$	523	(3.72)	$\frac{122}{94}$	216	(6.41)	739	(4.24)
Maternal Carry	$\frac{3}{25}$	28	(0.20)	$\frac{0}{2}$	2	(0.06)	30	(0.17)
Present for Groom	$\frac{33}{28}$	61	(0.44)	$\frac{7}{14}$	21	(0.62)	82	(0.47)
Touch	$\frac{24}{21}$	45	(0.32)	$\frac{6}{6}$	12	(0.36)	57	(0.32)
Social Investigate	$\frac{20}{9}$	29	(0.21)	$\frac{12}{6}$	18	(0.53)	47	(0.26)
Sex Present	$\frac{67}{62}$	129	(0.92)	$\frac{28}{27}$	55	(1.63)	184	(1.06)
Football Position	$\frac{10}{14}$	24	(0.17)	$\frac{9}{5}$	14	(0.42)	38	(0.22)
Genital Inspection	$\frac{29}{19}$	48	(0.34)	$\frac{5}{7}$	12	(0.36)	60	(0.34)
Climb On	$\frac{21}{10}$	31	(0.22)	$\frac{2}{1}$	3	(0.09)	34	(0.19)

Continued on following page

Behavior[a]	Frequency						Total	
	R-9			R-12				
Remain On	21/3	24	(0.17)	6/0	6	(0.18)	30	(0.17)
Climb Over	111/76	187	(1.33)	49/32	81	(2.40)	268	(1.54)
Lateral Approach	1493/1490	2983	(21.22)	631/619	1250	(37.11)	4233	(24.29)
Ankle Clasp	49/46	95	(0.68)	23/8	31	(0.92)	126	(0.72)
Thrust	33/30	63	(0.45)	18/6	24	(0.71)	87	(0.49)
Hip Touch	64/61	125	(0.89)	27/11	38	(1.13)	163	(0.94)
Bite	16/22	38	(0.27)	1/6	7	(0.21)	45	(0.26)
Chase	23/21	44	(0.31)	1/8	9	(0.27)	53	(0.30)
Charge	22/34	56	(0.40)	7/18	25	(0.74)	81	(0.46)
Threat	31/85	116	(0.83)	19/33	52	(1.54)	168	(0.96)
Avoid	437/270	707	(5.03)	143/66	209	(6.21)	916	(5.26)
Squeal		17	(0.12)		5	(0.15)	22	(0.13)
Approach	962/818	1780	(12.66)	28/12	40	(1.19)	1820	(10.44)
		14,059			3,368		17,427	

[a]See Smith (1977) for operational definitions.

[b]Percent of total for R-9.

[c]Percent of total for R-12.

[d]Percent of total (R-9 and R-12).

[e]$\frac{I}{R} = \frac{Initiates}{Receives}$

labels to these various clusters. Rather than use the traditional classification of types of play (i.e., approach-avoid,
rough-and-tumble), the five clusters have been designated play
classes (PI-V), and are represented as the first five clusters
on the dendogram. It seems unwise to assign traditional labels to these classes, for considerable subjective impression
exists as to which motor patterns should be included in the
traditional classes. Further, in some cases, these traditional labels would be misleading as to the types of behaviors
included in each class.

The use of the cluster analysis method allows the empirical categorization of behavioral elements based on an analysis
of causal similarities. However, the clusters of behavior,
although having similar causal mechanisms, do not allow for
unwarranted assumption of unitary causal forces. Each cluster
may, indeed, have its own underlying causal base, and in all
analyses are treated individually. As van Hooff (1973) noted,
these statistically defined clusters do not allow inference as
to the physiological basis of the underlying causal mechanisms, although cluster analysis allows a starting point for
investigations of this type (e.g., Jones, 1968; Maurus and
Pruscha, 1973).

The classification of social behavior patterns by nonhuman
primate researchers has typically been on a qualitative rather
than an empirical basis. The molar behavior categories, tra-
ditionally derived from functional considerations, most commonly recognized are, for example: contact aggression, non-
contact aggression, submission, sexual, grooming, other social
(Bernstein et al., 1974a), agonistic and non-agonistic (Kaufmann, 1967), aggression, sexual, anxiety, affinity, play
(Alexander and Harlow, 1965), and play, fear and aggression-
hostility (Chamove et al., 1972). Although these molar categories are essentially validated by this study, the lack of
empirical basis, other than face validity, for assigning these
labels by many researchers is troublesome. Table IV presents
a summary of the cluster play classes, with their component
behavior patterns. As can be seen, it is difficult to categorize these play classes into traditional social play classifications (e.g., contact, rough-and-tumble, approach-withdrawal).

Intuitively, one might expect other behaviors to be included in these clusters (e.g., push, grapple, wrestle, bounce
approach, bounce avoid), but since they were not linked with
existing clusters at the criterion point, they will not be
treated in this report. Admittedly, if the cluster analysis
had been performed on each age/sex class independently, some
variation in results may have been noted; however, comparisons
across age/sex classes would have been difficult.

TABLE IV. Five Play Classes (PI-PV), Their Behavioral Components, Frequency of Occurrence, and Mean Hourly Rate for All Subjects by Group

Play Class	Behavioral Components[a]	Frequency			Mean Hourly Rate		
		R-9	R-12	Total	R-9	R-12	Total
PI	take from, food display, head down, circle, jump over-contact, jump over-noncontact, slaps, slaps at	1884 (27.2)[b]	442 (25.4)	2326 (26.9)	15.70[c]	11.05[d]	13.38[e]
PII	jump on, pull/grab, tail pull, hold	1470 (21.3)	305 (17.5)	1775 (20.5)	12.23	7.62	9.92
PIII	mouth, roll onto back	1236 (17.9)	292 (16.8)	1528 (17.6)	10.38	7.30	8.84
PIV	crouch stare, head over shoulder, bounce charge, bounce chase, lateral bounce approach	1410 (20.4)	417 (23.9)	1827 (21.1)	11.64	10.42	11.03
PV	play face, head under	917 (13.3)	286 (16.4)	1203 (13.9)	7.68	7.15	7.42
TOTAL		6917	1742	8659			

[a] See Smith (1977) for operational definitions.
[b] Percent of total.
[c] Hourly rate (120 hours total – 5 hours/subject).
[d] Hourly rate (40 hours total – 2 hours/subject).
[e] Mean hourly rate for both groups combined.

IV. DISCUSSION

As Marler and Hamilton (1966) note, the description and classification of behavior may be the most important single issue in many ethological studies. The precise establishment of a systematic classification scheme of behavior patterns was first suggested by Jennings (1906), who referred to them as action systems.

A systematic catalogue of behavior patterns consists of physical descriptions (Hinde, 1966), which ideally should include all the details of the behavioral event. In actuality, this is an exceptionally difficult task, for an observer typically omits those characteristics of the behavioral event which are not important to him. As Hinde (1971:412) notes, "...we must therefore be constantly aware of the extent to which our categories really do represent discontinuities and the degree to which they are a matter of convenience." In any event, ethological studies should start with accurate descriptions of the behavioral repertoire of the species, the ethogram (Eibl-Eibesfeldt, 1975). However, the description of behavior patterns is only the first step in the classification of behavior.

Problems manifest themselves when the observer begins to classify relatively stereotyped behaviors into classes which are recognizable in different animals and in different encounters. In general, behavior can be classified as: (a) easily recognizable patterns, which are often only motor acts (e.g., bite, slap); or (b) sequences of movement which have a specified goal, "strategems" (e.g., attack, avoid, chase) (Poole, 1973). Clearly, the behavioral inventory used in this study is composed of a combination of motor acts and strategems. This combination, however, does not place a serious limitation on this study, but merely points out that often functional, rather than purely descriptive, labels are attached to behaviors, although caution in interpretation should be exercised when this procedure is employed.

Cluster analysis leads to a classification of behavior, which results in lumping of certain actions together, thereby revealing the integrated structure of the behavioral repertoire. The hierarchical nature of the structure is revealed through the cluster analysis. First, cluster analysis demonstrates that all elements represent a collection of individual behaviors and, as more individual types are classed together, higher and higher orders of hierarchical structures are revealed (van Hooff, 1973). The identification of the structure of the behavioral repertoire adds to our understanding of the integrated systems of the behaviors which are present.

Multivariate statistical techniques, especially cluster

analysis, reveal the general aspects of frequently occurring
phenomena; however, the specific acts of unique occurrences,
although possibly theoretically significant, find little expression in these statistical techniques. Certainly, when
considering the behavior of animals in a variety of ecological
contexts, the probability of finding such unique events increases, as the system of variables which affect behavior increases. The tendency for increasing complexity in the integration of behavior and the increasingly greater role of perception and higher integrative neural functions is present as
one moves toward more complex organisms. This tendency should
alert us to Rioch's (1967) and Mason's (1965) warnings that
attempts at quantification of behavior should not be at the
complete expense of the qualitative and anecdotal narrative,
which often is the only possible avenue to understanding these
unique behavioral events.

In sum, multivariate statistical analysis offers the possibility of quantifying behavioral observations in such a manner that underlying patterns or associations become apparent,
which otherwise might be obscured. Nonetheless, qualitative
observations are necessary to add a meaningful perspective to
these data.

A. *Social Play Defined*

Cluster analysis generated results which are intrinsically
interesting, but become more intriguing when the common, or
generally accepted definitions of social play are considered.
Rather than an undifferentiated uniform behavioral event, results indicate that in this experimental setting, social play
is organized into at least five different classes of behavior,
which occur as temporally distinct units. Total social play
comprises 49.20% of the total social interactions (n = 14,059),
and approximately 20.42% of the total behavioral observations
(n = 33,869) for the R-9 group.

Although a number of investigators have described social
play in rhesus macaques (i.e., Breuggeman, 1976; Harlow, 1959,
1962, 1964; Harlow et al., 1963; Hinde and Spencer-Booth,
1967; Lichstein, 1973a,b; Loy and Loy, 1974; Reynolds, 1972;
Rosenblum, 1961; Rosenblum et al., 1969), none have used multivariate statistical analyses to define in quantitative terms
the behavior patterns which comprise this behavioral category.
In the study of primate behavior, relatively few attempts have
been made to use multivariate statistics in this manner, save
the work of Altmann (1965, 1968), Chamove et al. (1972), Jensen et al. (1969) and Locke et al. (1964). Furthermore, cluster analysis, specifically, has found application only in the
work of Maurus and Pruscha (1973), Morgan et al. (1976),

Pruscha and Maurus (1973) and van Hooff (1970, 1973). None of
these investigators, however, has been specifically concerned
with social play. In this light, this study demonstrates that
such approaches can be fruitfully employed to understand the
variability of expression in social play. Although not a
"cure-all" for problems of classification of behavior, cluster
analysis presents, at least, an attractive alternative.

REFERENCES

Alexander, B. K., and Harlow, H. F. Social behavior of juve-
    nile rhesus monkeys subjected to different rearing condi-
    tions during the first six months of life. *Zoologische
    Jahrbucher Physiologie 71*, 489-508 (1965).
Altman, J. "Organic Foundations of Animal Behavior". Holt,
    Rinehart and Winston, New York, 530 pp. (1966).
Altmann, J. Observational study of behavior: sampling meth-
    ods. *Behaviour 49*, 227-265 (1974).
Altmann, S. A. A field study of the sociobiology of rhesus
    monkeys, *Macaca mulatta*. *Ann. N. Y. Acad. Sci. 102*, 338-
    435 (1962).
_____ Sociobiology of rhesus monkeys. II. Stochastics of
    social communication. *J. theoret. Biol. 8*, 490-522 (1965).
_____ Sociobiology of rhesus monkeys. IV. Testing Mason's
    hypothesis of sex differences in affective behavior.
    Behaviour 32, 49-69 (1968).
Anderson, C. O., and Mason, W. A. Early experience and com-
    plexity of social organization in groups of young rhesus
    monkeys *(Macaca mulatta)*. *J. comp. physiol. Psychol. 87*,
    681-690 (1974).
Anderson, T. W., and Goodman, L. Statistical inference about
    Markov chains. *Ann. math. Statist. 29*, 1112-1122 (1957).
Andersson, M. Temporal graphical analysis of behaviour se-
    quences. *Behaviour 51*, 38-48 (1974).
Andrew, R. J. The information potentially available in mam-
    malian displays, *in* "Non-verbal Communication" (R. A.
    Hinde, ed.), pp. 179-204. Cambridge University Press,
    Cambridge (1972).
Aschoff, J. Circadian rhythms in birds. *Int. orn. Congr. 14*,
    81-105 (1967).
Baerends, G. P. Aufbau des tierischen Verhaltens. *Handb.
    Zool. 8*, 1-32 (1956).
Baerends, G. P., and van der Cingel, N. A. On the phylogene-
    tic origin of the snap display in the common heron *(Cerdea
    cinera L.)*. *Symp. zool. Soc. Lond. 8*, 7-24 (1962).
Baerends, G. P., Drent, R. H., Glas, P., and Groenewold, H.
    An ethological analysis of incubation behaviour in the

herring gull. *Behaviour Suppl.* *17*, 135-235 (1970).

Bekoff, M. The development of social interaction, play, and metacommunication in mammals: an ethological perspective. *Q. Rev. Biol.* *47*, 412-434 (1972).

_____ Animal play and behavioral diversity. *Am. Nat.* *109*, 601-603 (1975).

_____ A sequence analysis of social interaction in infant canids: social play and aggression. Unpublished manuscript (n.d.).

Bernstein, I. S. Daily activity cycles and weather influences on a pigtail monkey group. *Folia primat.* *18*, 390-415 (1972).

_____ Activity patterns in a gelada monkey group. *Folia primat.* *23*, 50-71 (1975).

Bernstein, I. S., Gordon, T. P., and Rose, R. M. Aggression and social controls in rhesus monkey *(Macaca mulatta)* groups revealed in group formation studies. *Folia primat.* *21*, 81-107 (1974a).

_____ Factors influencing the expression of aggression during introductions to rhesus monkey groups, *in* "Primate Aggression, Territoriality, and Xenophobia" (R. L. Holloway, ed.), pp. 211-240. Academic Press, New York (1974b).

Bernstein, I. S., and Mason, W. A. Activity patterns of rhesus monkeys in a social group. *Anim. Behav.* *11*, 455-460 (1963).

Billingsley, P. "Statistical Inference for Markov Processes". University of Chicago Press, Chicago, 75 pp. (1961).

Blalock, H. M. "Causal Inferences in Nonexperimental Research". University of North Carolina Press, Chapel Hill, 200 pp. (1968).

_____ "Social Statistics". McGraw Hill, New York, 583 pp. (1972).

Breuggeman, J. A. "Adult Play Behavior and Its Occurrence Among Free-Ranging Rhesus Monkeys *(Macaca mulatta)*". Ph.D. Dissertation, Northwestern University, Evanston, Illinois, 123 pp. (1976).

Cane, V. Some ways of describing behaviour, *in* "Current Problems in Animal Behaviour" (W. H. Thorpe, and O. L. Zangwill, eds.), pp. 361-388. Cambridge University Press, Cambridge (1961).

Carpenter, C. R. Sexual behavior of free-ranging rhesus monkeys *(Macaca mulatta)*. *J. comp. Psychol.* *33*, 113-162 (1942).

Chamove, A. S., Eysenck, H. J., and Harlow, H. F. Personality in monkeys: factor analyses of rhesus social behavior. *Q. J. exp. Psychol.* *24*, 496-504 (1972).

Chance, M. R. A. Social structure of a colony of *Macaca mulatta*. *Br. J. Anim. Behav.* *4*, 1-13 (1956).

Chatfield, C. Statistical inference regarding Markov chain

models. *Appl. Statist. 22,* 7-20 (1973).

Chatfield, C., and Lemon, R. E. Analysing sequences of behavioural events. *J. theoret. Biol. 29,* 427-445 (1970).

Cole, A. J. (ed.) "Numerical Taxonomy". Academic Press, New York, 324 pp. (1969).

Cormack, R. M. A review of classification. *J. R. Statist. Soc. 134,* 321-367 (1971).

Delius, J. D. A stochastic analysis of the maintenance behaviour of skylarks. *Behaviour 33,* 137-178

Dingle, H. Aggressive behavior in stomatopods and the use of information theory in the analysis of animal communication, *in* "Behavior of Marine Mammals, Vol. 1, Invertebrates" (H. E. Winn, and B. L. Olla, eds.), pp. 126-156. Plenum Press, New York (1972).

Dolhinow, P. J., and Bishop, N. The development of motor skills and social relationships among primates through play. *Minnesota Symp. Child Psychol. 4,* 141-198 (1970).

Eibl-Eibesfeldt, I. "Ethology, the Biology of Behavior". Holt, Rinehart and Winston, New York, 625 pp. (1975).

Fagen, R. Selective and evolutionary aspects of animal play. *Amer. Nat. 108,* 850-858 (1974).

Fano, R. M. "Transmission of Information". M.I.T. Press, Cambridge, Massachusetts, 389 pp. (1961).

Fentress, J. C. Development and patterning of movement sequences in inbred mice, *in* "The Biology of Behavior" (J. A. Kiger, ed.), pp. 83-131. Oregon State University Press, Corvallis (1972).

Gilmore, J. B. Play: a special behavior, *in* "Current Research in Motivation" (R. N. Haber, ed.), pp. 343-354. Holt, Rinehart and Winston, New York (1966).

Gordon, T. P., and Bernstein, I. S. Seasonal variation in sexual behavior of all-male rhesus troops. *Am. J. phys. Anthrop. 38,* 221-226 (1973).

Haldane, J. B. S., and Spurway, H. A statistical analysis of communication in *Apis mellifera* and a comparison with communication in other animals. *Insectes Soc. 1,* 247-283 (1954).

Harlow, H. F. Love in infant monkeys. *Scient. Am. 200,* 68-74 (1959).

_____ The heterosexual affectional system in monkeys. *Am. Psychol. 17,* 1-9 (1962).

_____ Early social deprivation and later behavior in the monkey, *in* "Unfinished Tasks in the Behavioral Sciences" (A. Abrams, H. H. Garner, and J. E. P. Toman, eds.), pp. 154-173. Williams and Wilkins, Baltimore (1964).

Harlow, H. F., Harlow, M. K., and Hansen, E. W. The maternal effective system of rhesus monkeys, *in* "Maternal Behavior in Mammals" (H. L. Rheingold, ed.), pp. 254-281. Wiley, New York (1963).

Heiligenberg, W. Random processes describing the occurrence of behavioral patterns in a cichlid fish. *Anim. Behav.* *21*, 169-182 (1973).

Hinde, R. A. "Animal Behavior". McGraw-Hill, New York, 534 pp. (1966).

_____ Some problems in the study of the development of social behavior, *in* "The Biopsychology of Development" (E. Tobach, L. R. Aronson, and E. Shaw, eds.), pp. 411-432. Academic Press, New York (1971).

Hinde, R. A., and Rowell, T. E. Communication by postures and facial expressions in the rhesus monkey *(Macaca mulatta)*. *Proc. zool. Soc. Lond.* *138*, 1-21 (1962).

Hinde, R. A., and Spencer-Booth, Y. The behaviour of socially living rhesus monkeys in their first two and a half years. *Anim. Behav.* *15*, 169-196 (1967).

Hines, M. The development and regression of reflexes, postures, and progression in the young macaque. *Cont. Embryol.* *30*, 155-209 (1942).

Hopf, S. Study of spontaneous behavior in squirrel monkey groups: observation techniques, recording devices, numerical evaluation and reliability tests. *Folia primat.* *17*, 363-388 (1972).

Jardine, N., and Sibson, R. "Mathematical Taxonomy". John Wiley and Sons, London, 286 pp. (1971).

Jennings, H. S. "The Behavior of Lower Organisms". Columbia University Press, New York, 366 pp. (1906).

Jensen, G. D., Bobbitt, R. A., and Gordon, B. N. Studies of mother-infant interactions in monkeys *(Macaca nemestrina)*: hitting behavior. *Proc. 2nd Int. Congr. Primat.* *1*, 186-193 (1969).

Jones, N. G. B. Observations and experiments on causation of threat displays of the Great Tit *(Parus major)*. *Anim. Behav. Monogr.* *1*, 75-158 (1968).

Kaufmann, J. H. Social relations of adult males in a free-ranging band of rhesus monkeys, *in* "Social Communication Among Primates" (S. A. Altmann, ed.), pp. 73-98. University of Chicago Press, Chicago (1967).

Khinchin, A. I. "Mathematical Foundations of Information Theory". Dover, New York, 120 pp. (1957).

Kummer, H. Soziales Verhalten einer Mantelpaviangruppe. *Beit. schweiz. Z. Psychol.* *33*, 1-91 (1957).

Lichstein, L. "Play in Rhesus Monkeys: I) Definition. II) Diagnostic Significance". Ph.D. Dissertation, University of Wisconsin, Madison, 117 pp. (1973a).

_____ Play in rhesus monkeys: I) definition. II) diagnostic significance. *Diss. Abstr. Intl.* *33*, 3985 (1973b).

Locke, K. D., Locke, E. A., Morgan, G. A., and Zimmerman, R. R. Dimensions of social interactions among infant rhesus monkeys. *Psychol. Rep.* *15*, 339-349 (1964).

Loizos, C. Play in mammals. *Symp. zool. Soc. Lond. 18*, 1-9 (1966).

⸻ Play behavior in higher primates: a review, *in* "Primate Ethology" (D. Morris, ed.), pp. 176-218. Aldine, New York (1967).

Lorenz, K. Z. Gestaltwahrnehmung als Quelle wissenschaftlicher Erkenntnis. *A. exp. angew. Psychol. 4*, 118-165 (1959).

Loy, J., and Loy, K. Behavior of an all-juvenile group of rhesus monkeys. *Am. J. phys. Anthrop. 40*, 83-96 (1974).

Marler, P. Developments in the study of animal communication, *in* "Darwin's Biological Work" (P. R. Bell, ed.), pp. 150-206. Cambridge University Press, Cambridge (1959).

⸻ Communication in monkeys and apes, *in* "Primate Behavior: Field Studies of Monkeys and Apes" (I. DeVore, ed.), pp. 544-584. Holt, Rinehart and Winston, New York (1965).

Marler, P., and Hamilton, W. J. "Mechanisms of Animal Behavior". Wiley, New York, 771 pp. (1966).

Mason, W. A. The social development of monkeys and apes, *in* "Primate Behavior: Field Studies of Monkeys and Apes" (I. DeVore, ed.), pp. 514-543. Holt, Rinehart and Winston, New York (1965).

Maurus, M., and Pruscha, H. Classification of social signals in squirrel monkeys by means of cluster analysis. *Behaviour 47*, 106-128 (1973).

McQuitty, L. L. Similarity analysis by reciprocal pairs for discrete and continuous data. *Educ. psychol. Measur. 26*, 825-831 (1966).

Miller, S. Ends, means, and galumphing: some leitmotifs of play. *Am. Anthrop. 75*, 87-98 (1973).

Morgan, B. J. T., Simpson, M., Hanby, J. A., and Hall-Craggs, J. Visualizing interaction and sequential data in animal behaviour: theory and application of cluster analysis methods. *Behaviour 56*, 1-43 (1976).

Müller-Schwarze, D. Ludic behavior in young mammals, *in* "Brain Development and Behavior" (M. B. Sterman, D. J. McGinty, and A. M. Adinolfi, eds.), pp. 229-249. Academic Press, New York (1971).

Nelson, K. The temporal pattern of courtship in the glandulocaudine fishes *(Ostariophyai, Churacidae)*. *Behaviour 24*, 90-146 (1964).

Norton, S. On the discontinuous nature of behavior. *J. theoret. Biol. 11*, 229-243 (1968).

Oakley, F. B., and Reynolds, P. C. Differing responses to social play deprivation in two species of macaque, *in* "The Anthropological Study of Play: Problems and Perspectives" (D. F. Lancy, and B. A. Tindall, eds.), pp. 179-188. Leisure Press, Cornwall, New York (1976).

Orloci, L. An agglomerative method for classification of plant communities. *Ecology 55*, 193-205 (1967).

_____ Information analysis in phytosociology: partition, classification, and prediction. *J. theoret. Biol. 20,* 271-284 (1968).

Overall, J. E. Note on the scientific status of factors. *Psychol. Bull. 61,* 270-276 (1964).

Palmgren, P. On the diurnal rhythm of activity and rest in birds. *Ibis 91,* 561-576 (1950).

Plutchik, R. The study of social behavior in primates. *Folia primat. 5,* 70-79 (1964).

Poirier, F. E. Introduction, *in* "Primate Socialization" (F. E. Poirier, ed.), pp. 3-28. Random House, New York (1972).

Poirier, F. E., and Smith, E. O. Socializing functions of primate play. *Amer. Zool. 14,* 275-287 (1974).

Poole, T. B. Cine film and computer analysis of rapid sequences of behaviour. *Med. Biol. Illus. 23,* 170-175 (1973).

Poole, T. B., and Fish, J. An investigation of playful behaviour in *Rattus norvegicus* and *Mus musculus* (Mammalia). *J. zool. Soc. Lond. 175,* 61-71 (1975).

Pruscha, H., and Maurus, M. A statistical method for the classification of behavior units occurring in primate communication. *Behav. Biol. 9,* 511-516 (1973).

Reynolds, P. C. "Play and the Evolution of Language". Ph.D. Dissertation, Yale University, New Haven (1972).

Reynolds, V. Problems of non-compatibility of behaviour catalogues of a single species of primates, *in* "Contemporary Primatology" (S. Kondo, M. Kawai, and A. Ehara, eds.), pp. 280-286. Karger, Basel (1975).

Richter, C. P. Animal behavior and internal drives. *Q. Rev. Biol. 2,* 307-342 (1927).

Rioch, D. McK. Discussion of agonistic behavior, *in* "Social Communication Among Primates" (S. A. Altmann, ed.), pp. 115-122. University of Chicago Press, Chicago (1967).

Rosenblum, L. A. "The Development of Social Behavior in the Rhesus Monkey". Ph.D. Dissertation, University of Wisconsin, Madison, 134 pp. (1961).

Rosenblum, L. A., Kaufman, I. C., and Stynes, A. J. Interspecific variations in the effects of hunger on diurnally varying behavior elements in macaques. *Brain, Behav. Evol. 2,* 119-131 (1969).

Rowell, T. E. Hierarchy in the organization of a captive baboon group. *Anim. Behav. 14,* 430-443 (1966).

_____ Female reproductive cycles and the behavior of baboons and rhesus macaques, *in* "Social Communication Among Primates" (S. A. Altmann, ed.), pp. 15-32. University of Chicago Press, Chicago (1967).

_____ Intra-sexual behaviour and female reproductive cycles of baboons *(Papio anubis). Anim. Behav. 17,* 159-167 (1969).

Rowell, T. E., and Hinde, R. A. Vocal communication by the

rhesus monkey *(Macaca mulatta)*. *Proc. zool. Soc. Lond.* *138*, 279-294 (1962).

Sade, D. S. "Ontogeny of Social Relations in a Free-Ranging Group of Rhesus Monkeys". Ph.D. Dissertation, University of California, Berkeley, 185 pp. (1966).

Slater, P. J. B. Describing sequences of behavior, *in* "Perspectives in Ethology" (P. P. G. Bateson, and P. H. Klopfer, eds.), pp. 131-153. Plenum Press, New York (1973).

Slater, P. J. B., and Ollason, J. C. The temporal pattern of behaviour in isolated male zebra finches: transition analysis. *Behaviour 42*, 248-269 (1972).

Smith, E. O. "The Relationship of Age to Social Behavior in Pigtail Macaques *(Macaca nemestrina)*". M.A. Thesis, University of Georgia, Athens, 101 pp. (1972).

_____ "Social Play in Captive Rhesus Macaques *(Macaca mulatta)*". Ph.D. Dissertation, Ohio State University, Columbus, 235 pp. (1977).

Sokal, R. R., and Sneath, P. H. A. "Principles of Numerical Taxonomy". W. H. Freeman, San Francisco, 573 pp. (1973).

Steinberg, J. B., and Conant, R. C. An informational analysis of the inter-male behaviour of the grasshopper *(Chortophaga viridifasciata)*. *Anim. Behav. 22*, 617-627 (1974).

Symons, D. "Aggressive Play in a Free-Ranging Group of Rhesus Macaques *(Macaca mulatta)*". Ph.D. Dissertation, University of California, Berkeley, 213 pp. (1973).

Tinbergen, N. The objectivistic study of the innate behaviour of animals. *Biblthca. biotheor. 1*, 39-98 (1942).

_____ On aims and methods of ethology. *Z. Tierpsychol. 20*, 410-433 (1963).

van Hooff, J. A. R. A. M. A component analysis of the structure of the social behaviour of a semi-captive chimpanzee group. *Experientia 26*, 549-550 (1970).

_____ A structural analysis of the social behaviour of a semicaptive group of chimpanzees. *Europ. Monogr. soc. Psychol. 4*, 75-162 (1973).

von Cranach, M., and Frenz, H.-G. Systematische Beobachtung, *in* "Handbuch der Psychologie, Band 7: Sozialpsychologie, Halbband I: Theorien und Methoden" (C. F. Graumann, ed.), pp. 269-331. Verlag für Psychologie, Göttingen (1969).

Wells, G. P. Spontaneous activity cycles in polychaete worms. *Symp. Soc. exp. Biol. 4*, 127-142 (1950).

Wiepkema, P. R. An ethological analysis of the reproductive behavior of the bitterling *(Rhodeus amarus Bloch)*. *Archs. néerl. Zool. 14*, 103-199 (1961).

ETHOLOGICAL STUDIES OF PLAY BEHAVIOR
IN CAPTIVE GREAT APES[1]

Terry Maple
Evan L. Zucker

Department of Psychology and
Yerkes Regional Primate Research Center
Emory University
Atlanta, Georgia

I. INTRODUCTION

Nonhuman primates, like people, appear purposive, passion-
ate, and particularly playful. While there exist many detailed accounts of primate play, few studies have concerned
the great apes. Of these few, the vast majority are studies
of chimpanzees *(Pan troglodytes)*. In the wild (van Lawick-
Goodall, 1968) and in captivity (Savage and Malick, 1976), the
chimpanzee exhibits play from infancy through adulthood. As
both captive and field studies indicate, play is an important
factor in the normal social development of young apes. As
Loizos (1967) and Groos (1898) point out, the animal that
plays (practices) will become more proficient in its behaviors
and will, therefore, hold a selective advantage over less "ex-
pert" conspecifics.
     Of the field reports on apes, Goodall's (1965) is the most
complete for chimpanzees. Schaller's (1963) work on gorillas

[1]*This research is supported by the following sources: NIH
grant RR00169 to the Yerkes Regional Primate Research Center,
HD00208 to the Emory University Experimental Psychology Program, an NIH Biomedical Research Support Grant to Emory University, Faculty Research Grants from Emory University to Dr.
Terry Maple, and NSF Grant BMS 75-06287 to Dr. Ronald D. Nadler, whom we gratefully acknowledge as the principal senior
collaborator in much of our research.*

provides useful descriptions for comparison. However, field descriptions of orang-utan play (cf., Harrisson, 1962; MacKinnon, 1974) lack sufficient detail. Similarly, there exist no adequate descriptions of play behavior for the pygmy chimpanzee *(Pan paniscus)*.

Typical of great ape play studies in captivity is the work of Jacobsen, Jacobsen, and Yoshioka (1932) in which social play was broken down into play-threatening/attacking and swaggering postures. Other behaviors such as exploration, manipulation, and bodily acrobatics were seen as potentially social. Similarly, Bingham (1927) described chimpanzee play as composed of hand extension, grasping and tumbling, tickling, and chasing. From this report, some excellent notes on forms of parental play are also available for study.

Our research has been concerned with the acquisition of data on social behavior and social development in *Pan* (Clifton, 1976; Southworth and Maple, in preparation), *Pongo* (cf., Maple, in press; Zucker, Mitchell, and Maple, in press), and *Gorilla* (cf., Hoff, Nadler, and Maple, 1977; Wilson *et al.*, 1977). In this review, we will describe the basic components of play behavior in captive orang-utans and gorillas. We will also refer to data acquired on young chimpanzees in their interactions with orang-utan peers. It is our future objective to obtain a large data base from studies of each of the great ape species represented at the Yerkes Regional Primate Research Center. Using similar (if not identical) data acquisition techniques, acquired by the same (or similarly-trained) researchers under captive conditions, we are attempting to make useful comparisons of pongid social behavior. At present, our data are greater for the orang-utan than other species, and we will concentrate on this taxon within the context of this paper. Most writers have agreed that play be divided into two categories, *social* and *non-social*. In this paper, we will focus on the former, although we will describe the latter as it occurs in certain social or potentially social situations.

II. METHODS AND PROCEDURES

The subjects of our studies are described in Table I. In all, we have studied twenty orang-utans (ranging from infancy to 21 years of age), three juvenile chimpanzees, and seven gorillas, the latter currently residing in a heterosexual group. Our general procedures consisted of daily or three times weekly observation sessions of one to three hours in duration. Most observations were carried out in the morning between the hours of 0900 - 1100. Care was taken to keep

TABLE I. Subject Information

Name	Sex	Age[a]	Location[b]	Study	Parentage and Rearing
ORANG-UTANS					
Lipis	M	20	Zoo	Sociosexual Behavior/ Paternal Play	Feral born
Bukit	M	20	Zoo	Sociosexual/Proceptivity	Feral born
Sampit	M	21	YRPRC	Sociosexual	Feral born
Padang	M	18	YRPRC	Sociosexual	Feral born
Durian	M	18	YRPRC	Sociosexual	Feral born
Bagan	M	19	YRPRC	Sociosexual	Feral born
Dyak	M	19	YRPRC	Sociosexual	Feral born
Sungei	F	21	Zoo	Sociosexual/Infant Development	Feral born
Sibu	F	21	Zoo	Sociosexual/Proceptivity	Feral born
Lada	F	19	YRPRC	Sociosexual	Feral born
Paddi	F	18	YRPRC	Sociosexual	Feral born
Ini	F	18	YRPRC	Sociosexual	Feral born
Lunak	M	3.5-5.5	Zoo/YRPRC	Sexual Development/ Paternal Play	Sibu X Lipis; mother-reared
Merah	M	Birth-1	Zoo	Infant Development	Sungei X Lipis; mother-reared
Kanting	M	8-8.5	YRPRC	Sexual Development	Jowata X Sampit; nursery-reared
Loklok	M	7.5-8	YRPRC	Sexual Development	Datu X Tuan; nursery-reared
Biji	F	7	YRPRC	Sexual Development	Tupa X Dyak; nursery-reared

Continued on following page

Name	Sex	Age[a]	Location[b]	Study	Parentage and Rearing
Ayer	M	2.5	Play Area	Interspecies Play	Bali X Tuan; nursery-reared
Patpat	M	3.75	Play Area	Interspecies Play	Sungei X Lipis; nursery-reared
Teriang	M	3	Play Area	Interspecies Play	Paddi X Padang; nursery-reared
GORILLAS					
Rann	M	14	Field Station	Social Development	Feral born
Segou	F	14	Field Station	Social Development	Feral born
Choomba	F	14	Field Station	Social Development	Feral born
Shamba	F	18	Field Station	Social Development	Feral born
Akbar	M	Birth-1.5	Field Station	Social Development	Shamba X Rann or Calabar; mother-reared
Bom Bom	M	Birth-1.5	Field Station	Social Development	Segou X Rann; mother-reared
Machi	F	Birth-1.5	Field Station	Social Development	Choomba X Rann or Calabar; mother-reared
CHIMPANZEES					
Barbara	F	3	Play Area	Interspecies Play	Sonia X Hal; nursery-reared
Ellie	F	3.3	Play Area	Interspecies Play	Cookie X Homer; nursery-reared
Joice	F	3.75	Play Area	Interspecies Play	Flora X Hal; nursery-reared

[a] Ages given are ages at the time the study was conducted. For feral born animals, ages given are estimates.

[b] Zoo = Grant Park Zoo, Atlanta, GA; YRPRC = Yerkes Regional Primate Research Center.

daily observation periods consistent within a study, insofar as possible. Trained observers stood or sat within six feet of the respective enclosures and scored from 20 to 150 behaviors, depending on the full objectives of the study and the date the study was initiated. Our earlier attempts utilized a smaller list of behaviors, whereas we have been recently using a larger ethogram which we have constructed for orang-utans (cf., Maple, in press) and are in the process of constructing (with Nadler) for gorillas. We have attempted to record all social interactions within the cage and have used a coded, sequential behavior system for frequencies and durations of behavioral events. These efforts, particularly in the beginning, have been supplemented by the use of super-eight photography of highly complex social interactions, e.g., play, sexual behavior, and maternal care. Specific details of the studies described here are presented in the respective sections that follow.

The locations of our studies varied in the amount of available space and in the composition of the social and physical environment. The majority of our orang-utan studies have utilized adult heterosexual dyads, five of which were formed at the Yerkes Regional Primate Research Center and two of which were formed at the Grant Park Zoo in Atlanta, Georgia. The latter two pairs enjoyed a larger and more complex physical environment, but neither habitat was ideal. In one of the observed pairs, a newborn infant resided with them for the duration of the study. The gorillas are housed in a heterosexual group (one adult male, three adult females, and three infants) at the Yerkes Field Station in a large outdoor enclosure, an ideal captive habitat for gorillas. The orangutan - chimpanzee play study was conducted in the outdoor play area at the Yerkes Regional Primate Research Center on the Emory University campus.

III. ADULT MALE-OFFSPRING PLAY IN *Pongo*

Our first findings of great ape play were derived during an early study of social behavior in a captive orang-utan group (Zucker, Mitchell, and Maple, in press). In this study, we observed the interactions of the adult male, *Lipis*, his consorts, *Sungei* and *Sibu*, and *Sibu*'s four-year-old male offspring, *Lunak*. We were surprised at the high frequency of playful interaction between *Lipis* and the young male. Living at the Grant Park Zoo, the group had been on loan from the Yerkes Primate Center and were housed together for the duration of our three-month study until its breakup due to a

serious fight. From this research, we obtained 28 concise
descriptions of "paternal" play, 26 from films and two from
written notes (see Figure 1).

From these play bouts, ten principal behavioral components
were identified (Table II). The most frequent of these was
nonaggressive biting (mouth fighting), which occurred 39 times,
nearly 25% of the total number of behaviors. The second most
common behavior was *hand contact*, which occurred 34 times, and
appeared at least once in 57% of the interaction sequences.
The other frequent components were *dragging/pushing* (24), *following*, and *extremity* or *head investigation* (14 each). In

FIGURE 1.   *Adult male orang-utan* (Lipis) *rolling juvenile*
(Lunak) *during rough-and-tumble play bout. (Photo by E. L.
Zucker.)*

TABLE II. Behavioral Categories

Category	Definition
A. Hand contact	haptic contact with any part of the other animal's body. Includes swatting - brief contact with continued arm/hand movement.
B. Nonaggressive biting or mouth fighting	contact with other animal's body with open mouth, teeth visible.
C. Following	remaining proximate to the other animal as the other animal moved.
D. Dragging/pushing	movement, in contact with floor, towards or away from the other animal as a result of impetus applied by other animal. Includes rolling of other animal.
E. Extremity or head investigation	touching of other animal's appendages or head.
F. Mouth contact or oral exploration	contact with other animal with closed mouth or lips/tongue.
G. Hand extension	no contact made with other animal.
H. Hair pulling	common usage.
I. Falling	movement from a standing position to a sitting or prone position without impetus from another animal.
J. Face stroking	slow, repetitive vertical contact with face by fingertips of another animal.

decreasing frequency, the remaining behaviors were *mouth contact* (11), *hand extension* and *hair pulling* (7 each), *falling* (4), and *face stroking* (3).

In four of the 28 interactions, the initiator could not be clearly determined. However, in 20 of the remaining 24 cases, the four-year-old male *(Lunak)* initiated play. The adult male *(Lipis)* clearly initiated only four play bouts. Similarly, *Lunak* terminated sixteen of the interactions, while *Lipis* ended contact only twice. The mode of termination was generally withdrawal from proximity, whereby *Lunak* usually returned to his mother.

There were two common positions for these playful interactions: 1) both subjects on the floor, face-to-face; and 2) one or both animals hanging from the ceiling bars. On those occasions when only one partner was hanging, the other was reclining on the topmost platform in the cage. Both of these basic positions occurred with equal frequency, thirteen times each (43% of all play interactions). In the hanging position, *Lunak* hung the most frequently, accounting for twelve of the thirteen events. Less frequent positions were with *Lipis* on the floor and *Lunak* on the lower bench (twice) and both animals on the floor, face to back (once). As can be seen, social play in this pair reflects, in many instances (53%) the arboreal propensities of the orang-utan (see also Adult Heterosexual Play).

In examining the sequence of these recorded play components, nine of the 28 bouts began with *hand contact*, five with *hand extension*, and four with *head* or *extremity investigation*. Thus, contact by the hand of one or both subjects was the initial mode of interaction in 60% of the playful sequences, a not-too-surprising finding in light of the tactual skills of great apes. An inspection of the sequences revealed that the most commonly occurring pair of behaviors was *nonaggressive biting* following *hand contact*. After this pair in frequency was *dragging-pushing* followed by *nonaggressive biting*, and vice-versa. Other common pairs were *hand contact* followed by *nonaggressive biting*, and *nonaggressive biting* followed by itself.

Adult male-juvenile play in this situation may be characterized as rough-and-tumble. In this fashion, it differed from play between *Lunak* and his mother (*Sibu*). Although *Lunak* played roughly with *Sibu*, these interactions were generally unidirectional with *Sibu* tolerating play, but not reciprocating in kind.

The rather low rate of play initiation by the adult male orang-utan suggests that adult male orang-utans, although vigorous players, may not themselves seek out younger play partners with any great frequency. We are confident in labeling these interactions as playful, as fearful vocalizations were absent, behavior was reciprocal, neither animal attempted to flee and interactions were initiated and sustained by the smaller animal. In this latter category, our data differ from that of Jantschke (1972), who reported that the more dominant animals initiated play.

Regarding the motor patterns reported in this study for orang-utans, it is interesting to note that *hand extension* and *hand contact* are also integral parts of play for chimpanzees (cf., van Lawick-Goodall, 1968) and gorillas (Wilson *et al.*, 1977). *Hand contact* and *hand extension* may serve a metacommunicative function (cf., Altmann, 1967), indicating that whatever follows will be play rather than aggression.

IV. ADULT MALE-OFFSPRING PLAY IN GORILLA

     To compare to our data from the orang-utan, we refer now
to the paper of Wilson *et al*. (1977) completed in the course
of our collaboration with Nadler. These data are preliminary
and concern only one adult male gorilla interacting with three
offspring (2 male, 1 female) during their first year of life.
     The adult, silverbacked, male gorilla *(Rann)* exhibited an
active interest in the infants, particularly the two males.
Of the total contacts between the adult male and the infants,
68 out of 76 (87%) were initiated by the adult male. The percentage of contacts directed toward the two male infants differed significantly from that of the female infant (x^2 = 47.04,
p < .001), whereas the difference in this behavior between the
two male infants was not significant (x^2 = 2.53, p > .05).
Nearly all of the infant-initiated contact (98%) was from one
of the male infants, *Bom Bom*. While adult male contact with
both male infants was characteristically high, qualitative
differences were apparent. *Rann*'s interactions with one, *Ak-
bar*, were primarily playful. These play bouts typically involved the adult male touching, grasping, mouthing, and tickling the infant. A more detailed description of these and the
developing peer play interactions must await our further film
analysis (see Figure 2).
     The differences in play initiation between orang-utan and
gorilla males reflect species differences in social organization. Orang-utan males lead relatively solitary lives, whereas silverback gorilla males generally reside as members (or
leaders) of social groups (cf., Rodman, 1973; Schaller, 1963).
While it is unlikely that adult males of either species in the
natural habitat play as much as do our captive animals, the
potential for play is greater in the group-living gorillas
than in the solitary orang-utans. We are painfully aware,
however, that our sample size is much too small to strongly
support this conclusion. Further data are clearly required.

V. PEER PLAY

     Although we are just beginning to study peer play in in-
fant and juvenile gorillas and orang-utans (see Figures 3 and
4), we can report some preliminary data on conspecific interactions and a detailed description of play behavior between
juvenile orang-utans and chimpanzees.
     The first phase of a study of the initial responses to
non-maternal heterosexual contact in three juvenile orang-utan
males (two nursery- and one mother-reared) has recently been

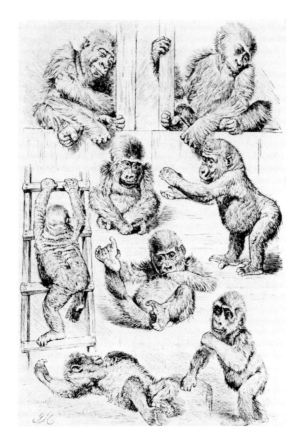

*FIGURE 2. Characteristic postures of young gorilla
(Brehm's* Tierleben, *1922; from Yerkes and Yerkes, 1929). Note
hand-clapping and foot clasping, basic solitary play behaviors which also occur during social interactions.*

completed. We hypothesized that the mother-reared male would
exhibit more appropriate sexual behavior than the two nursery-
reared males. In observations of their behavior together
(cf., Zucker *et al.*, 1977), the mother-reared male was the
only individual to exhibit pelvic thrusting toward his like-
sexed cage-mates. To our surprise, differences in sexual behavior were not apparent in the heterosexual condition, as all
three animals responded to the six-year-old (heterosexually
naive) female with vigorous play. It is quite possible that
orang-utans respond to all peers playfully until they have

reached adulthood, or some other threshold of sexual development. While it is too early in our research for us to say
that appropriate sexual behavior can be learned through play,
it is clear that captive mothers provide some early input by
mounting and *thrusting* against their babies at an early age
(cf., Maple, in press). We have observed no evidence of sex
differences in rough-and-tumble components. As is the case
with adult pairs, the sub-adult female played just as vigorously as did the three males with which she was housed.

VI. PLAY WITH ALIEN PEERS

Species differences in the behavior of captive-reared
juvenile chimpanzees and orang-utans in conspecific groups
were reported by Nadler and Braggio (1974). Our research was

FIGURE 3. A chase sequence in young gorillas. (Photo by
T. Maple.)

FIGURE 4.   Contact play in young gorillas.   Note the open-
mouth expression.   (Photo by T. Maple.)

designed to extend these findings to a mixed-taxa situation,
complementing previous work by the senior author on intertaxa
social behavior (cf., Maple, 1974; Maple and Westlund, 1975,
1977). Our first objective was to determine the propensity of
young apes to affiliate with members of another pongid species
of comparable age. Second, we wished to continue the development of our great ape ethograms by describing the motor patterns emitted during these interactions and by looking for species typical modes of interaction. We were surprised that the
vast majority of interactions between species were playful.
The subjects of this study were three juvenile orang-utan
males, and three juvenile chimpanzee females. We acquired our
data during six two-hour observation periods, during which time
we united the animals and recorded behavior through the use of
a super-eight movie camera and handwritten notes. One observer
filmed play interactions as they occurred (which is actually a
sample, since there were several playful interactions occurring
at once, and play was essentially continuous), while the other
observer recorded frequencies of both interspecific and intraspecific contact during fifteen-minute time samples. From the

film record, twenty-seven behavior categories were identified
(see Appendix I) from sixty-five discrete interspecific interactions.

A summary of 216 contacts initiated during the six fifteen-minute samples is presented in Table III. Of the 216
total contacts, 198 (91.7%) were interspecific, while only 18
(9 chimp-chimp; 9 orang-orang) were intraspecific. Of the 198
interspecific interactions, 128 were initiated by chimpanzees
and 70 were initiated by orang-utans ($X^2 = 17.18$, $p < .001$).

Figure 5 shows the frequency for the specific behaviors
that initiated interspecific interactions for 50 such interactions captured on film. For fifteen interactions, the
initiating behavior could not be determined. The most frequent behavior initiating play was *grabbing*, followed by *object stealing, hand contact, hair grasping/pulling, shoving/
pushing, hand extension*, and *jumping over other* (see Figure 6).

Table IV presents the total frequency of each behavior by
species during the 65 play bouts. The total number of behaviors emitted was not significantly different for the two species ($X^2_1 = .889$, $p > .5$), although the two species exhibited
significantly different behaviors ($X^2_{53} = 233.48$, $p < .001$).
Of the thirteen most frequent behaviors, chimpanzees emitted
more climbing, pushing/pulling, object stealing, jumping over

TABLE III. *Interspecific and Intraspecific Interactions
for Six Observation Periods, Broken Down by
Species, Initiator, and Recipient*

Initiator - Recipient	Observation Period						Total
	1	2	3	4	5	6	
INTERSPECIES							
Chimp - Orang	21	26	40	11	13	17	128
Orang - Chimp	18	2	8	9	16	17	70
							198
INTRASPECIES							
Chimp - Chimp	3	1	2	0	2	1	9
Orang - Orang	1	0	1	0	3	4	9
							18

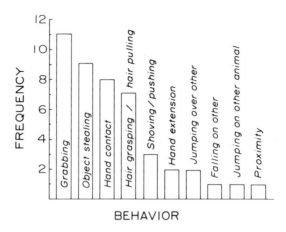

FIGURE 5. Frequencies of behaviors that initiated interactions between chimpanzees and orang-utans. (Adapted from
Clifton, 1976.)

FIGURE 6. Initiating a play bout. Note the chimpanzee's
terrestrial stance, and the orang-utan's grip on the climbing
apparatus. (Photo by T. Maple.)

other, and jumping to ground (Table V; U = 0, p $\leq$ .100).[2] Two
categories necessitate reciprocity, *mouth fighting* and *hand
grappling*, and there were no observed species differences in
the incidence of these behaviors. One behavior, *presents*, involved both a species and a sex difference, but this behavior
rarely occurred. The number of behaviors per interaction
ranged from one to sixty-six (M = 10.7, Md = 8).

The high degree of interspecific play which occurred during
this study is remarkable, as ample opportunity existed for
intraspecies play. It is not entirely surprising that young
chimpanzees and orang-utans should interact. For all their
differences, they are also quite similar in physical size, and
in many of their facial expressions. However, these similarities are not a prerequisite for interaction, since very diverse
primate species have been known to interact in an affiliative
manner (Maple, 1974).

Differences in the mode of play behavior observed in this
study reflect fundamental species differences. Chimpanzees
are primarily terrestrial animals with respect to play and
locomotor-type activity, and live in communities composed of a
number of small groups (van Lawick-Goodall, 1968). Orang-
utans, on the other hand, are principally arboreal apes, living
either as solitary animals (adult males) or in groups of two
or three, as in the case of females with their offspring
(Davenport, 1967; MacKinnon, 1974; Rodman, 1973). These differences in use of vertical space are retained in captive-born
animals (Nadler and Braggio, 1974). Behaviorally, chimpanzees
are very active, while orang-utans are lethargic, slow-moving,
and deliberate (Davenport, 1967; Yerkes and Yerkes, 1929).

Because of these activity differences, it was expected that
chimpanzees would engage in more active types of behaviors, and
would initiate more interactions because of their greater activity and gregariousness. The results supported these predictions, although both species initiated a large number of interactions.

One explanation for the high frequency of interspecific
interactions relative to intraspecific interactions is that
both chimpanzees and orang-utans responded to each other as
novel stimuli (cf., Baldwin and Baldwin, this volume). However, this tendency appears to have been remarkably resistant
to habituation since it did not wane with time. The novelty
of the situation may have been maintained by the fact that the

[2]*The remaining behaviors were too infrequent to compare.
In these analyses, Joice's behaviors were omitted due to her
extremely low behavioral output (see Page 129). Although not
statistically significant due to the small sample size, these
results were in the predicted direction.*

TABLE IV. Total Frequencies of Each Behavior for Each Species in the Interspecies Play Study[a]

Species														Behavior														Total
	1	2	3	4	5	6	7	8	9	10	11	12	13	14	15	16	17	18	19	20	21	22	23	24	25	26	27	
Chimpanzees	22	17	10	51	39	4	32	33	17	18	17	2	15	7	13	11	11	9	10	4	2	3	2	10	0	3	2	364
Orang-utans	80	41	49	26	5	31	5	38	17	5	1	0	15	4	0	2	0	0	0	4	2	0	0	10	3	1	0	339
Total	102	58	59	77	44	35	37	71	34	23	18	2	30	11	13	13	11	9	10	8	4	3	2	20	3	4	2	703

[a]Refer to Appendix I for the behaviors that correspond to the numbers.

TABLE V. Total Frequencies of Each Behavior for Each Subject in the Interspecies Play Study[a]

Subject														Behavior														Total
	1	2	3	4	5	6	7	8	9	10	11	12	13	14	15	16	17	18	19	20	21	22	23	24	25	26	27	
Barbara	11	14	5	33	21	1	21	22	7	13	12	0	5	3	8	5	7	4	10	3	1	2	0	4	0	0	1	213
Ellie	11	3	5	17	17	3	11	11	8	5	5	2	10	3	5	6	4	5	0	1	1	1	2	6	0	3	1	146
Joice	0	0	0	1	1	0	0	0	2	0	0	0	0	1	0	0	0	0	0	0	0	0	0	0	0	0	0	5
Ayer	21	14	26	7	1	6	1	9	0	0	0	0	5	0	0	0	0	0	0	2	0	0	0	3	0	1	0	96
Patpat	40	15	15	6	3	12	0	25	8	2	0	0	9	4	0	1	0	0	0	2	0	0	0	1	1	0	0	143
Teriang	19	13	8	13	1	13	4	4	9	3	1	0	1	0	0	1	0	0	0	0	2	0	0	6	2	0	0	100

[a]Refer to Appendix I for the behaviors that correspond to the numbers.

opportunity to interact was limited to only two hours/day. Extended exposure per day may have resulted in a decrease in interactions, and a decrease may also have been observed had the study continued for a longer period with repeated two-hour play sessions.

Alternatively, as our data indicate, an interaction with an animal different in its mode of play may have been reinforcing in some manner. Another reinforcing property of these play sessions may have been the opportunity to interact with an animal of roughly the same size. With the exception of *Joice* (a three-year-old female chimpanzee), the heavier animals interacted the most, the lighter animals the least. Nadler and Braggio (1974) also found body weight rather than age to be correlated with the number of interactions in conspecific groups of chimpanzees and orang-utans. We suspect that this subject's extremely low interactive rates and behavioral output may be due to brain damage incurred either prenatally or at parturition. *Joice* was a dizygotic twin, delivered by Caesarean section, and whose respiration needed to be stimulated pharmacologically. These birth traumas may be reflected in her social and play behaviors. On the other hand, *Teriang*'s high interactive rate can be explained by either the novelty or body weight hypothesis. Of all the animals, *Teriang* was the heaviest, but he was also different in appearance from the other two orang-utans. *Teriang*, a Bornean orang-utan, had a darker brown coat color than either of the other two orang-utans (Sumatran subspecies), as well as more developed cheek pads and throat sac.

Freeman and Alcock (1973) found that the majority of *intraspecific* social play of their young gorillas involved heterosexual interactions. In this study, interspecific interactions were exclusively heterosexual, but the only sexual behaviors that occurred were two *presents* by chimpanzees (*presents* are also chimpanzee submissive social signals). The absence of any other sexual behavior suggests that these *presents* were communicative in nature, rather than sexual.

It is possible that behavioral differentiation between the two species studied here is due to the sex of the animals (and we recognize the limitations of our sample), but these differences can be explained more parsimoniously as species differences. More climbing and jumping behaviors were observed in the chimpanzees than in the orang-utans, as well as more object stealing. The orang-utans, however, displayed more behaviors which involved greater use of the upper portions of their bodies, and less movement of their legs. It is surprising that they exhibited less climbing.[3] There was more hair pulling, grabbing and biting by the orang-utans than by the chimpanzees. The specific behaviors were often employed by each species to gain an advantage over the other (see Figures 7 and 8). Chimpanzees utilized their greater agility in play bouts. Often, a chimpanzee climbed or hung on the fence or climbing structures above an orang-utan's head until the orang-utan grabbed for it, then quickly withdrew out of the orang-utan's reach. Also, chimpanzees slowly climbed on the apparatus, and as soon as the orang-utan caught up, the chimpanzee dived through the bars and dropped to the ground. Chimpanzees were also observed to jump from an apparatus, and over an orang-utan by placing their hands on the orang-utan's shoulders and hopping as in "leap frog".

The orang-utan's strength compensated for the chimpanzee's agility and speed. Orang-utans often ignored a chimpanzee's presence only to suddenly lunge forward, grab a chimpanzee's arm, leg or hair, and bite. The orang-utan's strong hold prohibited the chimpanzee from escaping.

Several behaviors exhibited by chimpanzees were rare or absent among the orang-utan's behaviors. These included dangling, play-thrusting, walking backwards, walking sideways, jumping on other, presenting and hitting. Orang-utans exhibited a greater degree of mouth contact than did chimpanzees and repeatedly bit the face, hands, feet and back of chimpanzees during play. Nadler and Braggio (1974) reported a higher incidence of slapping and hitting behavior for chimpanzees than orang-utans. In this study and our study, chimpanzees spent more time on the ground than did orang-utans. The arboreal orang-utan's survival in its natural habitat is dependent upon strong grasping behaviors, making the orang-utan's mouth and lips important structures in interacting with their social and physical environments.

When orang-utans moved from one area to another, they usually moved around the periphery of the cage, holding the fence as they walked on the ground, or climbed around on the fence. Other captive nonhuman primates are known to remain near edges in their environments (Menzel, 1969). For the captive orang-utans in our study, however, their locomotion was highly dependent upon these boundaries, unlike the chimpanzees. In play interactions, orang-utans were frequently observed to lie on their backs, holding onto a climbing structure with their feet.

---

[3]The climbing apparatus was located in the central portions of the cage. Only one of the orang-utans spent its time in these inner portions. All three orang-utans were peripheral relative to the more mobile chimpanzees. This location may have been a factor in this difference.

FIGURE 7.   Appendage-biting by a young orang-utan.   (Photo by T. Maple.)

Even when not interacting, the orang-utans would keep at least one hand or foot attached to a permanent structure in their environment.

It was not always clear which behavior actually initiated a given play bout, since physical contact was not necessary. That is to say, subtle gestures cannot always be detected on film. Play bouts can be started by simply gaining proximity to another animal, and terminated by either moving out of proximity or by discontinuing behavioral output. Criteria must be established by which initiation and termination can be judged.

VII.   ADULT SOLITARY PLAY IN *Pongo*

Our studies of adult heterosexual pairs of orang-utans have been carried out in two different captive habitats (the Yerkes Regional Primate Research Center and the Grant Park Zoo in Atlanta).[4] Both solitary and social play are evident in these

[4]*Much of these data are discussed in a Senior Honor's Thesis by M. Beth Dennon (1977), who conducted most of the observations of adult pairs (Emory University Library).*

FIGURE 8. *Chimpanzee-orang-utan play. Note the orang-utan grasping and biting the chimpanzee's back, and the ambiguous facial expression of the chimpanzee. (Photo by T. Maple.)*

settings, although there are considerable individual differences in the amount of play exhibited. In solitary play, we have been concerned with self-motion play [*peragration* (Mears and Harlow, 1975)], rather than object play. While the latter has not been the focus of our work, we have noticed that both orang-utan and gorilla adults play with objects in similar ways. Both species place objects on their heads, slap their hands onto the ground while sitting and holding objects, pound rocks and sticks repeatedly, throw objects into the air, and roll around with objects in the same way that they roll with another animal (often emitting *play faces*). The self-motion play that we have seen in *Pongo* has been observed only in females. In female orang-utans, self-motion play is composed of *hanging* or *dangling* by the hands or feet, *somersaulting*, rapid *swinging* and *brachiation*, *spinning*, *sliding*, *dropping to the floor*, *water stomping*, and *pounding* the floor, all of which are typically accompanied by a *play face* and a vocalization which we call *tongue-gulps* (vocal clicks). In the seven adult pairs studied, solitary self-motion play was found to be characteristically brief in duration, one minute and twelve seconds on the average (range: 10 sec to 12-1/2 min). In three of the seven females, solitary play was especially prominent in their repertoire, in three others solitary play occurred, but infrequently, while in the remaining animal, solitary play was absent.

In agreement with MacKinnon (1974), our studies revealed that adult female solitary play can be followed by copulation. In fact, solitary play may be an indicator of female hormonal status, as we have found evidence in one female that it correlates with proceptive sexual behaviors (Maple, Zucker, and Dennon, in press). Residing in the zoo environment, an adult female emitted play behavior which consisted of cyclic periods of play overlapping the proceptive behaviors of *following, positioning* and *pelvic-thrusting* against the male. During the course of our research into orang-utan sexual behavior, the female exhibited gradual increases in solitary play spanning the days 12-32, 46-55 and 73-94. Solitary play was not observed in this animal after the 94th day, which coincided with the onset of pregnancy. An adult male showed interest in these activity periods and frequently approached the female during her solitary play bouts. As a result of this interest, social play followed on three occasions. In one case, *Bukit* chased *Sibu* after a solitary play bout and engaged in ventro-ventral copulation.

Another adult female also exhibited a cyclical trend in solitary play while paired with an adult male. These play peaks occurred during the days 21-44, 57-70 and 79-94. Coincident with these play peaks were increases in social grooming by this female, although the male did not appear to take interest in these activities; he did not reciprocate the grooming nor did he initiate any other social behaviors.

VIII. ADULT SOCIAL PLAY IN *Pongo*

In our observations, we have observed frequent social play in two pairs of captive orang-utans, less frequent play in four pairs, and no social play in one pair. During the course of these observations, ninety-three bouts of social play were recorded and all but seven were initiated by the females. Social play was initiated most commonly by the female dangling by her feet above the male (cf., also adult male-offspring play) and extending her hand to him. While in this position, females typically hit or slapped the male. The male's response was generally to pull the female to the floor after which a rough-and-tumble play bout occurred. An alternative mode of initiation occurred with the female in a sitting position. In these instances, the female rocked into contact with

the male with her head (which we call *head butting*). When
males initiated play, the initiating behaviors were the same as
those of females: *hand extension, head butting* and *dangling-
hitting*.

During the course of adult heterosexual play bouts, we
identified a number of motor patterns, such as *wrestling, rolling other, slapping, hitting, mouth fighting, biting hands/feet*
and *hair pulling*. These behaviors were consistent with other
descriptions of orang-utan social play (cf., Abel, 1818, cited
in Yerkes and Yerkes, 1929; MacKinnon, 1974; Yerkes and Yerkes,
1929; Zucker, Mitchell, and Maple, in press). It is interesting to note that in our study, thus far, copulation has never
been observed to follow a bout of *social* play.

The average duration of social play bouts in our study of
adults was three minutes, nineteen seconds. Scores ranged from
nine seconds to twenty-three minutes and twenty-five seconds.
To end a bout, females withdrew during play more frequently
than did the males. In 82 play bouts where withdrawal was recorded, 60 (73%) were initiated by females. Female-initiated
social play was cyclic for one adult female, who played at
highest rates on days 16, 46, 51 and days 73-93. As in the
case of solitary play, this activity was coincident with proceptive behaviors as previously described. During her pregnancy, female-initiated social play was not observed. No other
female's play was cyclic.

IX. PLAY SIGNALS

Since it occurs so often during chimpanzee social play,
and that of other species as well, the wide-open mouth with
covered or partially covered teeth (van Hooff, 1962) has become known as the "play-face" (see Figure 9). This expression
has also been observed in gorillas (see Figure 4) and orang-
utans. However, the presumed function of the play-face, in
our view, has not been precisely determined (indeed, it may
not be possible to do so). For example, in Loizos' review
(1967), she described a modified play-face as observed by
Goodall at the Gombe Research Station in Tanzania:
    "...it is possible that this represents a higher
    intensity of play....it is also possible that in cer-
    tain socially ambiguous situations the increased baring
    of the teeth (and particularly of the upper row) in the
    play-face represents the introduction of an element of
    appeasement since the facial expression now bears more
    resemblance to the grin face, which can have an appeas-
    ing function" (Loizos, 1967:205).

FIGURE 9.   Play-face?   (Photo by T. Maple.)

The play-face itself has also been described as an appease-
ment gesture, informing the other animal that its approach is
not an aggressive one.   In this sense, the play-face may fulfill the definition of metacommunication (cf., Altmann, 1967).
We have found the play-face to be an unreliable indicator of
play in chimpanzees, gorillas, and orang-utans for it may or
may not precede a vigorous play bout.   Moreover, because of the
certainty of play-biting, the play-face is difficult to distinguish from biting when the animals are in contact (see Figure 6).

To make matters worse, the play-face occurs as a transition
from vocalizations to grimaces, and is so transient as to be
useless during intensive play-contact.   Thus, we have found the
play-face to be of very limited use, in that its predictive
value has not yet been empirically verified.   That it often
occurs in connection with play, however, cannot be denied.   Its
utility and social function need to be further explored.

X.   CONCLUSIONS AND COMMENTARY

From our studies of a large number of orang-utans, we have
learned that the motor patterns and positions assumed during
play bouts in captivity are basically similar throughout the

life span of the animal. Both males and females exhibit rough-
and-tumble play, and adults play frequently with each other
and their offspring. Although not exhaustive, our data have
allowed us to generate an ethogram of play for the captive
orang-utan and to compare it to the behavioral repertoire of
other great apes. In many ways, orang-utan play resembles
gorilla and chimpanzee play, but there appear to be distinct
differences as well. Many of these may be explained in terms
of differences in morphological and social organizational adaptations. Since we are attempting to study the great apes under
similar conditions, we believe that our research will eventually yield valid comparisons of social behavior and social
development. This effort can only be properly understood in
relation to the data acquired from field studies of each respective species. Play behavior, in particular, is likely to
be overly represented in the repertoire of captive animals, but
we trust that its form and function are similar in captivity
and in the wild. Our studies are no substitute for data from
the wild, and we aim to render our data comparable to field
data. Indeed, it is our hope that members of our research
group will be able to study great apes in the wild to make
such comparisons themselves. In carrying this out, we follow
the advice of Robert Yerkes, who encouraged a laboratory-field
interchange in his early conception of primate studies (Yerkes,
1916).

Yerkes did not himself engage in direct comparative studies
of orang-utans, gorillas and chimpanzees, being restricted by
a population almost entirely composed of chimpanzees. The current Yerkes collection poses a challenge to the comparative
animal behaviorist in that ample numbers of subjects exist for
detailed study. Our research will be enhanced when we have
improved our captive habitats, and it is our aim, particularly
for the orang-utan, to develop captive environments which re-
semble as closely as is possible their natural habitat. We are
currently in need of better developmental data and are slowly
acquiring it (cf., Hoff et al., 1977). We are studying mother-
reared and nursery-reared great apes as they are born, but it
is slow work. With other primates as well, the captive environment creates some problems which must be overcome if sufficient numbers of animals are to be born and successfully
reared.

The functions of play are best understood when field and
laboratory workers get together to see when, where, and under
what conditions play is probable. Those of us who work under
captive conditions, in lab and zoo, look forward to further
exchanges.

XI. SUMMARY

1. Like the common chimpanzee, *Pan troglodytes* (van Lawick-Goodall, 1968), orang-utans and gorillas exhibit play behaviors throughout their life spans. With respect to frequency of play, there is a great amount of within-species variability, particularly among adults, although the behaviors themselves show a developmental constancy. The remaining great ape species, the pygmy chimpanzee, is currently being studied.

2. In captive groups, adult male orang-utans and gorillas interact in a rough-and-tumble manner with male juvenile and infant conspecifics, respectively. Behaviors directed towards immature females cannot be assessed at this time. Interspecies differences in rates of interactions can be attributed to differences in social organization, although further research of parental play is needed to assess the full range of potential. Mother-offspring play is characterized by tolerance on the part of the mothers, rather than active participation.

3. Orang-utan peer play, both intra- and interspecific, is also rough-and-tumble in style, independent of the sex of the play bout participants. When given the opportunity to interact with an alien species, both juvenile chimpanzees and orangutans preferred a member of the other species as a play partner. One explanation for this persistent effect was the novelty of the play situation. Alternatively, the opportunity to interact with a similar-sized animal, regardless of species, may have been reinforcing in some way(s).

4. Solitary play by social-living adult orang-utans and gorillas is frequently object-oriented. Self-motion play is also observed. However, in some instances, play by orang-utan females may be hormonally mediated, as evidenced by the cyclicity of play periods. Solitary play during these peak periods occasionally led to bouts of social play. Copulation has been observed to follow solitary play, but never social play. Social play bouts were usually terminated by the smaller animal (the female), as was the case with juvenile-adult male play.

5. Regardless of age, similar behaviors and styles of play appear in the repertoire of each species, reflecting the particular species' adaptations to its natural environments. Arboreal orang-utans are typically slow-moving and cautious, utilizing their strong upper bodies in play interactions. Orang-utans are also oral players, frequently biting and mouthing their partners. Chimpanzee behaviors were more terrestrial, chimpanzees being more active players. Chimpanzee mobility, however, was counteracted by orang-utan strength in interspecies play bouts. Gorilla play behaviors appear somewhat intermediate to the other two species, although as of now data are only preliminary, and await further research and analysis.

6. The "play-face" has been observed in the three species studied, although it has questionable *predictive* value as a signal. The wide-open mouth with covered or partially-covered teeth is also a transition expression seen just prior to a bite, a behavior prominent in great ape play.

7. By studying play developmentally, comparatively, and in a variety of captive environments, conjectures about possible functions can be made. However, small sample sizes and slow-to-mature animals make these approaches difficult and tedious. The extensive and plastic behavioral and communicative reper-toires of these apes make comparisons with non-anthropoid pri-mates a cautious endeavor. Furthermore, more detailed descrip-tions of play in free-ranging animals are needed to corroborate captive studies. By continuing captive investigations of play behaviors, the apes' potentials for play can be assessed.

ACKNOWLEDGMENTS

The authors acknowledge the assistance, support, and coop-eration of the following institutions and colleagues: Zoologi-cal Society of Atlanta, Inc., Atlanta Zoological Park, Dr. G. H. Bourne, Dr. L. D. Byrd, Ms. L. D. Clifton, Ms. M. B. Dennon, Mr. S. Dobbs, Mr. M. P. Hoff, Ms. S. Puleo, Mr. J. Roberts, Mr. M. E. Wilson, and Ms. S. Wilson.

REFERENCES

Abel, C. "Narrative of a Journey in the Interior of China, and of a Voyage to and from That Country, in the Years 1816 and 1817". Longman and Co., London (1818). *Cited in* Yerkes, R. M., and Yerkes, A. W. "The Great Apes. A Study of Anthropoid Life". Yale University Press, New Haven (1929).

Altmann, S. A. The structure of primate social communication, *in* "Social Communication Among Primates" (S. A. Altmann, ed.), pp. 325-362. University of Chicago Press, Chicago (1967).

Bingham, H. C. Parental play of chimpanzees. *J. Mammal. 8,* 77-89 (1927).

Clifton, L. D. Interspecific social behaviors of infant chim-panzees and infant orangutans in captivity. Unpublished Honor's Thesis. Emory University, Atlanta (1976).

Davenport, R. K., Jr. The orang-utan in Sabah. *Folia prim-atol. 5,* 247-263 (1967).

Dennon, M. B. Affiliation, affection, and attraction in

captive adult pairs of orang-utans. Unpublished Honor's Thesis. Emory University, Atlanta (1977).

Freeman, H. E., and Alcock, J. Play behaviour of a mixed group of juvenile gorillas and orang-utans. *Int. Zoo Yrbk. 13,* 189-194 (1973).

Goodall, J. Chimpanzees of the Gombe Stream Reserve, *in* "Primate Behavior: Field Studies of Monkeys and Apes" (I. DeVore, ed.), pp. 425-473. Holt, Rinehart and Winston, New York (1965).

Groos, K. "The Play of Animals". D. Appleton and Co., New York (1898).

Harrisson, B. "Orang-utan". Collins, London (1962).

Hoff, M. P., Nadler, R. D., and Maple, T. The development of social play in a captive group of gorillas. Paper presented at the Inaugural Meeting of the American Society of Primatologists, Seattle, Washington (1977).

Jacobsen, C., Jacobsen, M., and Yoshioka, J. Development of an infant chimpanzee during her first year. *Comp. Psychol. Monogr. 9,* 1-94 (1932).

Jantschke, F. "Orang-utans in Zoologischen Garten". R. Piper and Co., Munchen (1972).

Loizos, C. Play behaviour in higher primates: a review, *in* "Primate Ethology" (D. Morris, ed.), pp. 176-218. Aldine, Chicago (1967).

MacKinnon, J. The behaviour and ecology of wild orang-utans *(Pongo pygmaeus). Anim. Behav. 22,* 3-74 (1974).

Maple, T. "Basic Studies of Interspecies Attachment Behavior". Ph.D. Dissertation, University of California, Davis (1974).

_____ Sexual behavior, sexual development, and sexual roles in captive orang-utans, *in* "Papers of the Harry F. Harlow Symposium" (S. J. Suomi, and L. A. Rosenblum, eds.). Van Nostrand Rheinhold, New York (in press).

Maple, T., and Westlund, B. The integration of social interactions between cebus and spider monkeys in captivity. *Appl. Anim. Ethol. 1,* 305-308 (1975).

_____ Interspecies dyadic attachment before and after group experience. *Primates 18,* 379-386 (1977).

Maple, T., Zucker, E. L., and Dennon, M. B. Cyclic proceptivity in a captive female orang-utan *(Pongo pygmaeus abelii). Behavioural Processes* (in press).

Mears, C. E., and Harlow, H. F. Play: early and eternal. *Proc. Nat. Acad. Sci. 72,* 1878-1882 (1975).

Menzel, E. W., Jr. Naturalistic and experimental approaches to primate behavior, *in* "Naturalistic Viewpoints in Psychological Research" (E. P. Willems, and H. L. Rausch, eds.), pp. 78-121. Holt, Rinehart and Winston, New York (1969).

Nadler, R. D., and Braggio, J. T. Sex and species differences in captive-reared juvenile chimpanzees and orang-utans. *J. hum. Evol. 3,* 541-550 (1974).

Rodman, P. S. Population composition and adaptive organization among orang-utans of the Kutai preserve, *in* "Comparative Ecology and Behavior of Primates" (R. P. Michael, and J. H. Crook, eds.), pp. 171-209. Academic Press, New York (1973).

Savage, E. S., and Malick, C. Play and socio-sexual behavior in a captive chimpanzee *(Pan troglodytes)* group. *Behaviour 60,* 179-194 (1977).

Schaller, G. B. "The Mountain Gorilla: Ecology and Behavior". University of Chicago Press, Chicago (1963).

Southworth, K., and Maple, T. Peri-pubertal social behavior in captive pygmy chimpanzees *(Pan paniscus).* (in preparation).

van Hooff, J. A. R. A. M. Facial expressions in higher primates. *Symp. Zool. Soc. Lond. 8,* 97-125 (1962).

van Lawick-Goodall, J. The behaviour of free-living chimpanzees in the Gombe Stream Reserve. *Anim. Behav. Monogr. 1,* 161-311 (1968).

Wilson, M. E., Maple, T., Nadler, R. D., Hoff, M. P., and Zucker, E. L. Characteristics of paternal behavior in captive orang-utans *(Pongo pygmaeus abelii)* and lowland gorillas *(Gorilla gorilla gorilla).* Paper presented at the Inaugural Meeting, American Society of Primatologists, Seattle, Washington (1977).

Yerkes, R. M. Provision for the study of monkeys and apes. *Science 43,* 231-234 (1916).

Yerkes, R. M., and Yerkes, A. W. "The Great Apes. A Study of Anthropoid Life". Yale University Press, New Haven (1929).

Zucker, E. L., Mitchell, G., and Maple, T. Adult male-offspring play interactions within a captive group of orang-utans *(Pongo pygmaeus). Primates* (in press).

Zucker, E. L., Wilson, M. E., Wilson, S. F., and Maple, T. The development of sexual behavior in infant and juvenile male orang-utans. Paper presented at the Inaugural Meeting, American Society of Primatologists, Seattle, Washington (1977).

APPENDIX I.   BEHAVIORAL CATEGORIES IN INTERSPECIES PLAY

1. Grabbing: Rapid movement to hold other animal by putting
    arms around other or gripping other with hands.
2. Hair grasping/hair pulling: Common usage.
3. Non-aggressive biting: Contact with other's body with
    open mouth, teeth visible.
4. Climbing up apparatus: Climbing, jumping, and/or swinging
    on apparatus.
5. Pushing/pulling: Movement to free self from other animal.
6. Hand extension: Reaching toward another animal without
    contact.
7. Jumping to ground: Jumping or dropping to ground from a
    climbing structure, fence, or table.
8. Mouth fighting: Reciprocal contact with another's face
    with an open mouth.
9. Hand contact: Haptic contact with any part of another
    animal's body.
10. Object stealing: Grabbing or pulling an object away from
    another animal.
11. Jumping over other: Jumping/swinging around or over an-
    other animal.
12. Walks sideways: Takes steps to the side while watching
    another animal.
13. Hand grappling: Reciprocal grabbing and pulling at the
    hands of another animal.
14. Falling on other: Falling on another animal without im-
    petus supplied by another animal.
15. Dangling: Holding onto a structure with arms or legs
    hanging unattached to the apparatus.
16. Shoving/pushing: Forceful movement against another ani-
    mal. Includes shoving, running or swinging into
    other, and pushing.
17. Hitting: Slapping with open hand or hitting with fist.
18. Play thrusting: Slight back and forth motion of body while
    standing quadrupedally.
19. Jumping on other: Jumping up and down *on* another animal.
20. Swinging around bars: Grasping and swinging around ver-
    tical bars in a circular motion.
21. Grooming: Picking at hair of another animal.
22. Jumping near other: Jumping up and down *near* another
    animal.
23. Walking backwards: Walks backwards dragging hands and
    arms in front of body.
24. Following: Remaining proximate (arm's length away or
    less) as the other animal locomotes.
25. Mouth contact: Touching another animal's body with a
    closed mouth, or with lips and/or tongue.

26. Slapping ground: Slaps hands against ground or table.
27. Presents: Stands quadrupedally with ano-genital region
      towards another animal.

FUNCTIONS OF PRIMATE PLAY BEHAVIOR

Frank E. Poirier
Anna Bellisari
Linda Haines

Department of Anthropology
The Ohio State University
Columbus, Ohio

*"Play has the status of a religion. It is a system of be-
liefs with many sectarian views" (I. S. Bernstein, personal
communication). This view was communicated to the authors at
the symposium from which papers for the present volume were
drawn. The authors agree with Dr. Bernstein's tongue-in-cheek
assessment. As this paper will show, there are many tenets of
faith for play behavior, the cornerstone being that play is
functional.*

I. INTRODUCTION

As play behavior becomes the focus of continued research,
theories suggesting a single function for this complicated and
variable behavior seem less appropriate. The functions of
play have been discussed by many authors and, as more are suggested, the question arises of which one or ones are most important. We will look at some of these suggested play functions from an evolutionary perspective. Logically, natural
selection would favor a behavior that is adaptive in more than
one respect - being, as it were, "doubly adaptive". We suggest that play is an adaptive behavior both for the group and
the individual, although these adaptive functions sometimes
overlap. For example, it is adaptive for the group that the
individuals form an integrated unit in which communication effectively disseminates information. At the same time, it is
individually adaptive to learn this communication matrix to

participate in the group. We will show how play facilitates
both social integration and socialization.

There seem to be types of play, e.g. solitary play, which
function almost solely as a means of individual adaptation.
We will consider solitary play from the perspective of the individual and suggest its place in adaptation.

We will also consider some variables that influence play.
Age, sex, and habitat appear to affect both the quality and
quantity of play in most nonhuman primates.

Play is of primary importance in the maintenance of peer
integration and the socialization process. This paper will
attempt to show how this pattern emerged in primate evolution
and how it functions in present populations.

II. THE EVOLUTIONARY PERSPECTIVE

A. *Effectance Motivation*

Since play is a general mammalian characteristic, it is
reasonable to assume that this behavior had adaptive value
during mammalian evolution. The nature and adaptive value of
play becomes clearer when its magnitude and distribution are
considered. There is variability in type and complexity of
play within Mammalia and there is variability in learning capacity. Those mammals that have evolved the greatest capacity
for learning are also those that play most often (Poirier and
Smith, 1974).

Mammals depend heavily on trial and error learning (Chevalier-Skolnikoff and Poirier, 1977; Poirier, 1972b, 1973b).
Most learning among animals is for the purpose of satisfying
the primary needs of hunger, reproduction and self-preservation. However, a new behavioral trait has appeared among mammals which does not satisfy these immediate needs. This new
element, effectance motivation, is the tendency to investigate
and explore the environment without the immediate objective of
satisfying a prime biological function (Campbell, 1966; White,
1959).

Effectance motivation in young mammals is typified by the
playful and exploratory interaction of the individual with
those aspects of the environment providing changing feedback.
Play and exploration lead individuals to discover how the environment can be manipulated and changed, and increased knowledge about the environment increases chances for survival
(Campbell, 1966).

Effectance motivation is correlated with increasing complexity and gross size change in the neural structure. In
mammals, there is increasing size and complexity of the

cerebral hemispheres. With increasing size and neural complexity, information input becomes the limiting factor in
optimum learning, input being maintained through highly developed sense organs and effectance motivation. According to
Baldwin and Baldwin (1977), optimal levels of sensory stimulation are prime reinforcers that activate or motivate play.
Experiences provided by play and exploratory behavior are assimilated by the brain and utilized to predict the likely
course of external events and fit occurrences into the total
life experience. Through natural selection, play has evolved
the potentiality for several important functional roles in
primate development.

B. Prolonged Immaturity and Learning

Except for some nocturnal species, most primates live in
social groups. Washburn and Hamburg (1965) have suggested
that group life is a sociological response to the primate biological pattern of prolonged immaturity. (It is, of course,
possible that the situation is reversed; i.e., prolonged immaturity may be a biological response to group life.) Washburn and Hamburg (1965) argue that prolongation of preadult
life is biologically expensive, and that a major compensation
is learning. Despite restraints imposed upon the social
order, the long infancy period is advantageous, "...it provides the species with the capacity to learn the behavioral
requirements for adapting to a wide variety of environmental
conditions" (Washburn and Hamburg, 1965:620).
The trend toward prolonged immaturity is associated with
the increasing importance of learned behavior. The elongation
of the juvenile developmental period appeared early in primate
evolution, perhaps during the Eocene or Oligocene geological
epochs (Poirier, 1977). Schultz (1956) has shown a progres-
sive increase in preadult life as one moves from prosimians to
humans. The juvenile period of lemurs is approximately two
years, of macaques four years, of chimpanzees eight years, and
among modern *Homo* the juvenile period extends to sixteen
years.
Delayed maturation, especially among apes and humans, is
probably associated with less encephalization at birth. Since
hominoids are more vulnerable to predation and death during
the early years than are other animals with relatively greater
encephalization at birth, there must be some strong compensation for this developmental disadvantage. Possibly this compensation is learned behavior. During the longer growth period, individual experiences perform a subtle yet important
role in shaping behavior into a large number of effective
patterns (Bekoff, 1977). Prolonged immaturity enhances the

amount and complexity of learning possible, while increasing
opportunities to shape behavior to meet various local habitat
and social conditions. Flexibility of behavioral patterns
may be one important result of the longer dependency period
(Poirier, 1969a,c, 1970, 1972c, 1973b). There appears to be a
positive correlation between prolonged postnatal dependency
and increasing complexities of adult behavior and social relationships (Poirier and Smith, 1974).

The basis for the flexibility of nonhuman primate behavior,
the ability to adjust to new situations (e.g., Jay, 1968;
Poirier, 1969a,c, 1973b, 1974), may be related to play behavior. Play may have assumed a major role in the evolution of
"non-biologically programmed" responses to environmental
stresses. Washburn and Hamburg (1965) suggest that play leads
to a diversified sampling of the environment that is important
in habitat adaptation. Frequent variation and modified repetition of behaviors, plus the possibility for innovative behaviors, are essential to the plasticity of primate behavior.
Through play, as well as other forms of social behavior, more
time is provided for peer (Baldwin and Baldwin, 1977) and
adult contact. This may promote the socialization process and
help integrate youngsters into the social group. Fedigan
(1972:363) suggests that "...play is a type of social and solitary behavior which provides those qualities during the period in which the immature individual is developing species-
specific behavior, group-specific behavior, and social perception."

Play appears crucial to the learning processes of slowly
maturing primates. In those species with the longest period
of immaturity, e.g. chimpanzees, play is prolonged and varied.
This may ultimately lead to increased and more types of learning experiences.

An important correlate of increased and varied types of
learning is a more complex brain, the storage area for learned
information. It is of interest that Diamond and Hall (1969)
specify the mammalian association cortex as the neocortex
subdivision where prime evolutionary advancements have occurred, although Radinsky (1975) and others disagree. Although
primate learning skills are not solely accountable by neocortex volume relative to total brain volume, it is significant
that the primate neocortex is proportionately larger than that
of carnivores and rodents (Harman, 1957). Complex cognitive
processes and advanced learning skills are accommodated by increased cortical fissuration, increased numbers of cortical
units in the cortex fine structure, and the refinement of the
subcortical structure interrelating the thalamus and cortex
(Norback and Moskowitz, 1963; Rumbaugh, 1970).

Many studies have shown the importance of an enriched environment for the development of intelligence. Obviously, the

many varied activities involved in play behavior and the many
experiences gained in play become available in this enriched
environment. In fact, play itself provides a stimulating environment. Play is a behavior that allows or causes the environment to be stimulating. The environment is not the stimulus for development, but rather the interaction of the animal
with the environment is important. Laboratory experiments
show that infants raised in sensory-rich environments tend to
learn more exploratory skills and habituate to higher levels
of sensory input than infants reared in sensory-impoverished
environments (Harlow, 1969; Mason, 1968; Rosenblum, 1971).

Primates are social animals whose sociability is related
to their learning behavior. Much of this learning occurs during the period of prolonged immaturity, itself a concomitant
to group life. The primate brain facilitates learning - it
readily occurs. Primates learn to be social, and under normal
conditions such learning inevitably occurs (Washburn and Hamburg, 1965). Presumably, in most higher mammalian social systems, and particularly among primates, individual behaviors
are controlled by a continuous process of social learning
arising from group interactional patterns. The selective
value of learning to act according to social modes is obvious:
animals whose behavior does not conform to group norms are
less likely to reproduce, and may be ejected from the social
group. Social selection of this type apparently has a strong
stabilizing influence upon the genetic basis of temperamental
traits and motivational thresholds (Crook, 1970).

Social facilitation and observational learning seem to be
important vehicles for role assumption (Hall, 1965; Hall and
Goswell, 1964). It is in the context of early play experiences that youngsters learn their social roles and the ways
of the social group. Since animals must learn how to exist
within the social group on their own, perhaps as soon as they
leave their mothers, it comes as no surprise that most play
behavior occurs among youngsters.

Certain key aspects of social development are completed
early in life. Loy and Loy (1974) note that over eighty-five
percent of the communicative behaviors witnessed among adult
rhesus [based on Altmann's (1965) ethogram of the Cayo Santiago population] are found in twenty-month-old juveniles. The
absence of some behaviors was attributed to physical immaturity. This correlation between biological and behavioral
maturation is discussed by Chevalier-Skolnikoff (1973) in
stumptail macaques. [For a further discussion, see Bekoff
(1977), Chevalier-Skolnikoff (1977), Gibson (1977) and Parker
(1977)]. Loy and Loy's (1974) data also suggest that young
juveniles learn their roles and status within the group by an
early age; the socialization process is virtually completed by
the end of the juvenile period.

We have noted the evolutionary relationship between prolonged immaturity, neural development, learning and play behavior. Group life is an important adaptive strategy of many
nonhuman primates. Play functions to promote social cohesion
in several ways and on several levels - that of the group and
the individual.

III. FUNCTIONS OF PLAY: VALUES FOR THE INDIVIDUAL AND THE
      GROUP

A. *Overview*

Washburn (1973) notes that if the field observer listed
the kinds of daily behaviors witnessed according to the amount
of time they consume, the usual order would be: sleeping, obtaining food, eating, playing, resting, and social contact.
The amount of time animals spend playing clearly indicates
that play is adaptive.

The adaptive value of play behavior has been the subject
of much discussion. Two especially valuable sources of information are Baldwin and Baldwin (1977), listing thirty separate
functions of play, and Welker (1971), providing a similar list.
The Baldwins point to learning as the single most important
function of play behavior, subsuming all thirty of the various
functions. We suggest that the learning function of play has
three basic objectives: 1) learning motor skills through physical exercise in normal growth and development, 2) developing
perceptual skills and enhancing neural development through
sensory input for central nervous system stimulation, and 3)
enhancing individual and social learning experiences through
contact with members of the social group, resulting in social
integration, learning of certain behaviors, learning the environment, socialization of the individual to the species-
specific behavioral repertoire, predator escape, personality
development and learning gender roles.

Simonds (1974:200) aptly notes that "all the usual theories about the functions of play relate play to future goals
rather than to immediate needs." Future goals are difficult
to anticipate without longitudinal field studies; therefore,
efforts should be made to determine the adaptive value of play
behavior for the fulfillment of immediate needs of the developing individual at that stage of its development when it is
engaging in play. Play behavior is functional on two separate
levels - that of the individual and that of the entire social
group. The play group provides an individual with a context
to practice physical skills, to experiment with its environment, to develop its cognitive skills, and to learn its social

roles. This occurs with minimal danger of injury or harm,
since its playmates are usually equal in stature, and strength,
and protective adults are nearby. Play facilitates learning
of the group communication matrix, including tension releasing
and relaxing mechanisms. The adaptive values of play to the
entire social group include the social integrational function
of play (i.e., through play with peers a developing individual
becomes an effective member of the group). Also important is
innovative behavior which is frequently engaged in by youngsters during play, resulting in new traditions for the entire
troop (Burton and Bick, 1971; Itani and Nishimura, 1973).
Thus, play can be considered adaptive behavior from the individual or group perspective.

B. *Separation from Mother*

While prolonged immaturity and the necessary infant depen-
dence on its mother are very important for survival, the infant must make the transition to a self-sufficient adult.
Play is one of the first non-mother directed activities. As
Bekoff (1972) notes, peer play can ameliorate the effects of
maternal separation and make it a less traumatic experience
(Harlow, 1969; Tisza, Hurwitz and Angoff, 1970). Harlow, Harlow and Suomi (1971), studying the social recovery of isolate
monkeys, stated that the most critical and valid measures of
social recovery were social contact and play behavior. Fedigan and Fedigan (1977) note an interesting situation in which
peers apparently helped compensate for severe defects. In
this case, a severely handicapped youngster made a short-lived
reversal in its abnormal social life when a female age-mate
showed interest in him, seeking out the handicapped youngster
and engaging him in gentle play. This period of increased
social activity and peer interest lasted for a month and was
probably due to the female's attention. When she no longer
sought out the handicapped youngster, he returned to his pattern of mother-centered activity.

The Fedigans' work shows the results of a lack of development away from self and self-and-mother stages toward the peer
and group orientation. While normal infants showed this progression, the handicapped youngster did not. Using the readiness concept borrowed from human development, the Fedigans
wonder whether the serious motor and sensory deficiencies of
the youngster arrested his development at an early stage. His
inability to engage in normal play and exploratory behaviors
and his lack of communicatory skills were evident in everyday
behavior.

The above example suggests that social interaction can
ameliorate rather serious physical defects. Although the long

term effect of the social "therapy" cannot be predicted, play
helps the infant separate from its mother. If something in-
terrupts this normal separation of the intense mother-infant
bond, the young animal will not become a normal group member.

C. Social Communication

As the infant moves away from its mother, skills required
for group living become essential (Poirier and Smith, 1974;
Smith, 1973; Symons, 1973). As an integral part of the practice of adult social roles, play serves to fully acquaint an
animal with its species-specific, and perhaps group-specific,
communication matrix (Poirier, 1977). Socially deprived animals have problems with response integration and communication.
Such animals do exhibit most components of social behavior;
however, these are not combined into an integrated pattern and
are not effectively utilized in social interactions. Mason
(1963) believes this is due to a deficiency in sensory-motor
learning or "shaping". Although all basic postures, gestures,
and vocalizations are probably genetically coded, their effective combination and use in social interaction are dependent
upon experience. This applies to the sending as well as the
receiving of messages, because messages are only effective if
understood.

Most of what is considered communication in nonhuman primates consists of movements and facial expressions, as well as
vocalizations. These behaviors occur during play. A most
useful approach to understanding how communication skills are
acquired is to consider play a kind of grammatical structure
(Chomsky, 1965). During ontogeny, players learn the behavioral syntax as a mathematical game (Kalmus, 1969). During
play, there are probably rules which a youngster learns. The
acquisition of an adequate performance and competence in rules
of play is a developmental process. The rules a youngster
learns are not lacking some sort of logical connection or
structure. Altmann's (1965) stochastic analysis of rhesus
communication shows considerable predictability within the
communication system of properly socialized animals. Berkson's (1977) and Fedigan and Fedigan's (1977) data on physically and socially handicapped infants demonstrate the results
of an inability to learn these rules.

As a youngster develops more elaborate play behavior during maturation, it may order the rules of the game into the
correct sequence for proper functioning in a social unit
[Bekoff (1974) and Smith (1977) raise doubts about this]. The
key to the acquisition of these rules is in the sequencing of
playful interactions and the association of relatively dis-
junctive units of behavior into larger functional categories.

Adult behaviors may be learned through repetition in the play
context, making an understanding of the complex repertoire of
signals employed in play an area of fruitful research (Goyer,
1970; Poirier and Smith, 1974).

Further discussion of the progression to orderliness found
in play behavior occurs in the Baldwins' (1977) article. As
they note, play sequences are often a jumble of disorganized,
clumsy, fragmented and repetitious behaviors. This description well fits the playing of young animals; as animals mature
their play becomes more stereotyped.

The Baldwins argue that older players are on extinction
procedures, and this difference produces visible variation between the play patterns of young and old. Symons (1973) notes
that the play behavior of older animals is slower, less vigorous, has fewer movements, is less random, and contains minimal
chasing. In play behavior, we find a reduction of activities
as the animal matures, suggesting that with age comes an ordering of behavioral patterns and communication processes,
which is related to development and maturation in the central
nervous system (Chevalier-Skolnikoff, 1973, 1977; Gibson,
1977). Play may function to order, as well as to practice
communication skills. An increase in effective communication
may be related to a decrease in the amount of play. Alarm
calls, once part of a play mode, are now likely responded to
by the group because of their transference to the adult communication system.

D. Social Integration

Learning the communication matrix is one part of the more
complicated process of social integration that is essential
for successful group functioning. Social play facilitates
and maintains integration through continual contact with other
group members. Adults rarely play with each other, perhaps
partly because their social relationships are already well
established. When adults do play, it is usually with an infant or juvenile. This suggests that social play performs an
integrating role for young animals.

Etkin (1967) suggests that play may be a means of reiter-
ating stimulus exchange whereby social animals maintain familiarity with one another. During play, youngsters learn their
place in the group and develop in-group feelings. Play maintains pair relations in social mammals and brings individuals
into close social contact that often involves considerable
tactile contact. "Repeated play, day after day, enables the
regular playmates to maintain in group familiarity and habitu-
al bonds that persist beyond infancy and is usually reinforced

in later life by other processes of familiarization, for example, grooming..." (Simonds, 1974:201).

Playing animals learn patterns of social cooperation without exceeding certain limits of aggression. This cooperation
brings its own rewards; overly aggressive, non-cooperative
animals may be socially rejected and perhaps excluded from the
group (Diamond, 1970). During play, an animal may learn to
limit its aggression and to cooperate socially so that eventually it will achieve some degree of reproductive success.
Without knowledge of social conventions, the animal considerably diminishes its relative fitness. Hall (1965) suggests
that play among patas males may contribute toward the decision
of whether they stay within the group or are temporarily excluded.

Play behavior occurs primarily with one's peers; in this
context, one sees most clearly one of the functions of play -
development of normative social behavior. Mason's work (1963)
clearly indicates that animals with restricted social experiences (those raised in isolation) show strikingly abnormal
sexual, grooming, and aggressive social patterns. Laboratory
research has long shown that the full development of an animal's potentialities requires the stimulus and direction of
social forces normally encountered in the peer play group, as
well as in the larger social group (Harlow, 1963, 1969; Mason,
1963). Harlow's (1969) deprivation studies show that peer
play interaction may be even more important for the development of normal social behaviors than maternal interaction.
Even brief daily play sessions between infants raised with
surrogate mothers compensated for the absence of real mothering. Surrogate-raised infants allowed twenty minutes of play
per day with the peer group were considerably better adjusted
than infants raised with their mothers alone.

One of the most important mechanisms in the integration of
the social group is the dominance hierarchy. The basis of the
adult hierarchy may be formed in the play group (Poirier and
Smith, 1974, among others). Play may help youngsters find
their place in the social order (Baldwin and Baldwin, 1977;
Carpenter, 1934); however, Symons (1973) would disagree.
Through trial and error, the constant repetition of play behaviors, an infant may learn the limits of its self-assertive
capabilities. During play, youngsters compete for many items
such as food, sleeping positions and convenient pathways.
Early dominance patterns may appear during such activities as
play-wrestling and rough-and-tumble play. Wrestling bouts,
characteristic of much of male play, give a growing animal
practice in behaviors ultimately influencing its social position. Although the dominance structure among juveniles is
primarily a function of the individual's size and the mother's
status, during play youngsters gain social experience and

become familiar with dominant and subordinate roles (Dolhinow and Bishop, 1970). Thus, they learn behaviors that will most likely characterize their relations with others for much of their lives.

A "test" of the integrative function of play occurred when three so-called bachelor males joined a bisexual Nilgiri langur troop (Poirier, 1969a, 1970). Play behavior between an infant 2 and members of the bachelor group assumed a major role facilitating the latter's integration into the bisexual group. Play accounted for approximately thirty-one percent of the observed interactions during the first two days of merging. Play was particularly striking in this instance since the dominant male of the bachelor trio played with the infant. Once the males merged with the group, the dominant male ignored the infant. Once the males indicated non-aggression through playful interactions with the infant, they were accepted as members by the female core of the group.

Play, of course, is not the only group cohesive mechanism. Grooming is very important for maintaining cohesion. The Baldwins' (1977) findings show that, at least for short periods of time, groups can maintain social integration without play. In contrast to the Baldwins' data, however, Nilgiri langur troops in which play and grooming were minimal were loosely structured and prone to fission (Poirier, 1969a, 1973a). Thus, play, while not an absolute necessity for maintaining group cohesion, is an effective means for facilitating integration.

IV. SOLITARY AND NONSOCIAL PLAY

Besides performing social integrative functions adaptive for the entire group, play is also valuable for the individual. In solitary play behavior, the emphasis is on individual rather than group adaptation. The observable behavior associated with solitary play is not much different from social play. According to Mason (1965:529), "The distinction between social and nonsocial play is not a fundamental one. Characteristics that affect responses to inanimate objects, such as complexity, size, or mobility of the stimulus, also influence reactions to social stimuli." However, as Baldwin and Baldwin (1977) note, social games offer more potential variability than nonsocial games because of the greater number of combinations possible when an individual interacts with others rather than with the inanimate environment or with itself. Simonds (1974) restricts nonsocial play to activities such as manipulation of familiar objects and locomotor play, and he suggests that such activities are practice and experience in motor behavioral acts. "Such exercise ensures that the individual has the skill to escape a predator. Otherwise, the endurance and speed in running and climbing are unnecessary in day-to-day activities" (Simonds, 1974:201).

Nonsocial or solitary play may be preparation for certain kinds of future behaviors. For example, McGrew (1977) discusses manipulation of vegetation among chimpanzees as a precursor to tool use. Various studies have noted that youngsters are more prone to engage in exploratory manipulation than adults. Menzel (1969) studied the exploratory behavior of a corral group of chimpanzees and suggested that the decreased amount of apparent exploration in adults relates to the fact that they can visually acquire a good deal of information about the environment. They may also be able to explore the environment several times to gather the same kind and amount of information adults visually acquire (Jolly, 1972).

Young chimpanzees practice behaviors, in what has been described as nonsocial play, in preparation for adult life. van Lawick-Goodall (1967) describes incipient nest building among chimpanzee youngsters. "The motor experience here is not necessary for later life...but it is one way of developing skills at a time when mistakes will not be dangerous or disastrous" (Simonds, 1974:201).

Solitary play is adaptive for the individual, as it allows the animal to learn more about the environment, its own physical skills and the required species-specific behaviors outside the social realm, e.g., nest building, predator escape, etc.

V. VARIABLES INFLUENCING PLAY BEHAVIOR

A. *Gender and Role Learning*

As with any social behavior, several variables affect the amount, kind and duration of play behavior. Among these variables are age, gender and habitat.[1] As one of the key factors

[1]*References to these variables are: age (Dolhinow and Bishop, 1970; Frisch, 1968; Itani, 1958; Kawai, 1965; Kummer, 1971; Poirier, 1969a, 1970, 1973b, 1977; Poirier and Smith, 1974; Tsumori, Kawai and Motoyishi, 1965); gender (Burton, 1972; DeVore and Eimerl, 1965; Dolhinow and Bishop, 1970; Fedigan, 1972; Goodall, 1965; Hall, 1965; Harlow and Harlow, 1961; Knudson, 1971; Kummer, 1968; Lancaster, 1972; Lindburg, 1971; Poirier, 1971, 1972a,b, 1977; Poirier and Smith, 1974; Ransom and Rowell, 1972; Smith, 1972, 1977; Sorenson, 1970); habitat (Baldwin and Baldwin, 1973; Jolly, 1972; Rowell, 1966; Washburn and DeVore, 1961). Perhaps the best overall discussion of this topic is found in Baldwin and Baldwin (1977).*

affecting play behavior, gender has been discussed by many authors. It is clear that males play harder, begin play earlier and cease play at a later age. There are qualitative and quantitative differences in the play behavior of male and female primates (Hinde and Spencer-Booth, 1967). Harlow and Harlow (1966) distinguished the play of male and female laboratory rhesus at about two months. Similar differences occur among feral baboons (DeVore and Eimerl, 1965), cynocephalus and hamadryas baboons (Kummer, 1968; Ransom and Rowell, 1972), and vervets (Fedigan, 1972). Among human children, the frequency of rough-and-tumble play for males is significantly higher than for females (Knudson, 1971, 1973).

Dolhinow and Bishop (1970) suggest that a powerful endocrine effect influences gender differences in play. Females exposed to androgens early in development become masculinized as pseudohermaphrodites (Goy and Phoenix, 1972; Goy and Resko, 1972; Phoenix, Goy and Resko, 1968), and develop play patterns approximately intermediate between typical male and female patterns. Simonds (1974) also suggests that gender differences in play are influenced by hormonal levels. In regard to adults, he queries (p. 200), "...how do changes in the hormonal or other physiological systems eliminate the need for play?" That answer is not yet available.

In their extensive review of play behavior, Baldwin and Baldwin (1977) suggest other factors affecting differential play expression according to gender. Males are usually larger and stronger, a factor which may explain some differences in play activities. This, plus the endocrine influence, increases the likelihood that females experience more aversive contact during exploration or social play than males. Females of many species also reach socio-sexual maturity prior to males, hastening the termination of their play behavior. Poirier (1968a, 1970, 1972a,b, 1973b) argues that females are less explorative than males and suggests an adaptive function - they are less likely to risk their infants' lives. This difference may appear during early developmental stages as females drop out of play groups. The Baldwins (1977) suggest that the females' proclivity for less physical activity shapes their quiet, withdrawn, and gentle activities and their tendency to orient to object manipulation. This is supported by McGrew's (1977) observations on chimpanzee tool use and Tsumori's (1967) observation that females learn to attend to lower but safer arousal levels afforded by manipulatory play and exploration.

As females withdraw from the play groups, they may partake in "play mothering", a pattern reported by Jay (1968) and Poirier (1969a, 1970) for langurs, by van Lawick-Goodall (1967) for chimpanzees, by Lancaster (1971, 1972) for vervets, and by Baldwin (1969) for squirrel monkeys. The opportunity

for juvenile females to handle infants is useful practice for
adult female behavior. "It is possible that the care and
handling of infants by juvenile female vervets may be looked
upon as a variety of play behavior, one that is very similar
to the human pattern of young girls playing with dolls or infant siblings" (Lancaster, 1972:102).

The female's gentle play with an infant is reinforced by
her being allowed to retain the infant. If the female plays
quietly, she may experience many mild, novel play patterns
with it, and is unlikely to be hurt by the infant's mother.
Baldwin and Baldwin (1977) state that these contingencies
shape restrained, mother-like behavior during play mothering.
The skills learned by the female during this behavior are most
useful to her as a future mother.

Young females maintain close ties with adult females
learning the mothering role (Poirier, 1972a,c, 1973a,b, 1974,
1977), but young males remain in proximity to their peers.
Perhaps it is more important for males, especially subordi-
nates, to have a stable relationship with one another, and
with older, more dominant males than with females. Juvenile
females seem to develop their social relations during long
grooming bouts with other females and while holding and ex-
changing infants (Poirier, 1977). Peer play seems more important for males and provides an opportunity for many diversified interactions with those individuals with whom they will
eventually interact and contend. Male juveniles have been
known to leave the natal group and play with males of other
groups (Struhsaker, 1967). Unlike females, who in most cases
spend their entire lives in the natal group, males often leave
or are driven away. Perhaps intertroop play can facilitate
movement to a new group.

B. *Age*

Age is another readily definable variable influencing play
behavior; play decreases in frequency as an animal matures.
Adult play is an infrequent activity, especially among males.
Dolhinow and Bishop (1970) suggest that adults do not partake
in play because: 1) of the potential danger to a would-be
player whereby a misinterpreted signal could lead to more harm
than in juvenile play bouts, 2) adults generally recognize a
certain social space which is not conducive to close interactions and physical contact characteristic of play, and 3) activities without observable reason are less frequent among
post-adolescents (a tautological argument). Another explanation for the gradual decrease of play is the animal's inability to find the play novel and stimulating. Welker (1971) and
Mason (1963) discuss the optimal range for play stimuli.

Possibly, stimuli novel to the juvenile are no longer interesting to the adult. Stimuli that might elicit play in the
young sometimes acquire other functions for the adult. For
example, a play attack may no longer be interpreted as such
when the attacking animal is an adult.

Studies of Japanese macaques (Frisch, 1968; Itani, 1958;
Tsumori, 1967) and Nilgiri langurs (Poirier, 1968a, 1969b,
1970, 1972b, 1973b) suggest that immature animals are the most
explorative, flexible and innovative in behavioral patterns.
Immature individuals may test the environment more than the
adults because they are more open to change. Japanese macaque
data indicate that animals over three years of age stagnate in
their adaptability to new situations (Itani, 1958; Kawai,
1965), although on another test Tsumori (1967) found the critical age to be six or seven. Menzel's (1966) study of Japanese macaques confirms the general impression that age is an
important variable in new object manipulation and exploration.
"It is the playful infants and juveniles of any species who
approach a new object and who are most expendable if it turns
out to be dangerous. Conversely, the older the animal the
less likely it is to involve itself in an unknown situation..." (Jolly, 1972:348).

Kummer (1971) suggests a selective advantage for the behavioral conservatism of adults. A primate group has invested
a great deal of energy and experience in each adult member,
while investment in a young individual is much smaller. Re-
tention of traditions by inflexible adults provides a measure
of safety in the event that new behavior patterns prove to be
nonadaptive. Youngsters are the most obvious candidates for
acquiring new behavioral patterns since they are more easily
replaced. A species' survival may ultimately depend on allowing youngsters playful experimentation and environmental
manipulation. New behavioral patterns acquired during play,
new foods tasted, new travel routes navigated, may become crucial factors in the group's eventual adaptive success.

C. Habitat

The last variable to be discussed is habitat. Since play
is often cited as a group integrating behavior, it should be
expected that there is variation between terrestrial species
which require a tighter social organization and arboreal species with their looser structure (Sussman, 1977). Jolly
(1972) suggests that the amount of exploratory play and the
distance the infant is allowed to stray from mother reflect
the terrestrial/arboreal dichotomy. The environment influences early play patterns and reinforces the already present
social structure (Chalmers, 1972).

VI. SUGGESTIONS FOR FURTHER STUDY

We would like to suggest that it is appropriate to further test the functions of play behavior. There are many problems inherent in such testing, but then there is also the problem of simply defining play. Many functions of play have already been tested; others, to our knowledge, have not yet received proper attention.

Deprivation studies focus attention on some of the pitfalls of hypothesis testing. In these studies, animals were deprived not only of play stimuli, but the experimental design also denied them total social experiences. The resulting behaviors were then not simply a function of play deprivation, but of social deprivation generally (Baldwin and Baldwin, 1977; Bekoff, 1977).

A. Play and Learning

1. It has been argued that play is important in learning about the environment. Do animals in more complex habitats play more than those in so-called "simpler" habitats? How do habitat differences affect play behaviors; what is the relationship? Although a certain amount of behavioral flexibility is found in a species as a whole, each group's particular environment determines the amount of flexibility that is adaptive. For example, in a "difficult" environment, feeding takes more time; thus, flexibility is not as important. It should be expected then that as the difficulty of the environment increases, behaviors facilitating flexibility (such as play) would decrease (Baldwin and Baldwin, 1977). In the long run, this may be maladaptive. If play is habitat dependent, when the habitat is changed, there should be a change in both play and the social organization over time.

2. Play is often cited as a learning experience; however, there are few experimental studies testing what kind(s) of information is learned in play. If play facilitates learning, then species with a greater learning capacity should spend more time playing than those less dependent upon learning.

3. In what ways, if at all, does play teach an animal to act according to accepted social modes? Do animals that do not play also learn appropriate group behaviors? Can the same kind(s) of learning occur solely through observational learning without active participation in play behavior?

4. Many have stated that play facilitates the learning of the group communication matrix (Baldwin and Baldwin, 1977; Bekoff, 1972; Poirier and Smith, 1973; Symons, 1973). Can animals deprived of play, but animals that have observed group

life (for example, from an adjacent cage) learn the appropriate communication modes? Can communication skills be learned
without playing? If play is a series of jumbled behaviors
which become organized into recognizable patterns by adulthood, then a young player should show more variability of pattern than older animals (Symons, 1973).

   5. It has been suggested that play helps an animal learn
its place in the dominance hierarchy (Baldwin and Baldwin,
1977; Carpenter, 1934; Diamond, 1970; Dolhinow and Bishop,
1970; Jolly, 1972; Poirier, 1972a, 1974; Poirier and Smith,
1974). Recently, Symons (1973) has sought to disprove this
hypothesis. Has the final word been written?

   6. Can "maleness" and "femaleness" be further manipulated
in order to separate factors differentiating male and female
play? Hormonal factors, growth factors, adult roles, and
others have been suggested as explanations for differences in
male and female play (Baldwin and Baldwin, 1977; Dolhinow and
Bishop, 1970; Goy and Phoenix, 1972; Goy and Resko, 1972;
Harlow and Harlow, 1966; Hinde and Spencer-Booth, 1967;
Simonds, 1974). Are these factors exclusive of one another;
can others be defined?

   7. It has been suggested that play may be more important
for males than for females. Males are said to develop their
social relationships in play while females develop their social relationships during social grooming. This has yet to
be demonstrated. A concentrated study of the amount of time
young males and females play and groom might be useful. Is
the adult female social structure independent of the amount of
play as immature animals?

   8. How does play-mothering affect a female's ultimate
mothering behavior? If play mothering facilitates role learning, then the female not allowed contact with infants should
show less skill in caring for her first infant (Baldwin and
Baldwin, 1977; Lancaster, 1971, 1972; Poirier, 1972a, 1973a,
1977; Poirier and Smith, 1974).

B. Play and Maturation

   9. It has been argued that one reason older animals cease
playing is because stimuli activating play in youngsters are
no longer interesting to older animals. What types of stimuli
do animals of different ages attend to?

   10. Are various kinds of play more important at different
periods of development? Harlow (1969) and Poirier (1970) have
defined gross stages in the play of rhesus and Nilgiri langurs
respectively. Closer attention to the types of play in each
developmental stage might help clarify play functions.

   11. Are there gross neural differences between animals that

play and those that do not? Since we have argued that larger
brains, such as those found among primates, require information input, what happens if information input is restricted?
Would anatomical studies point to a relationship between play
behavior and neural development in slowly maturing animals?
If play enhances neural development and increases perceptual
skills, then an animal denied play should be expected to show
deficiencies in these areas.

12. If play facilitates separation from mother, then an
animal denied play should show increased time spent with
mother as compared with normal playing peers (Berkson, 1977;
Fedigan and Fedigan, 1977).

C. Stimulation Levels

13. Play behavior occurs most readily within certain optimum stimulation levels (Baldwin and Baldwin, 1977). What are
the stimulation levels for various species? Do the slowly
maturing forms, such as chimpanzees, have higher stimulation
levels and, thus, a greater need for play than more rapidly
maturing species?

D. Group Integration

14. If, in two groups of the same species, one was found to
be without play, there should be some other difference between
the groups.

15. Play is often cited as a group integrating mechanism;
animals that play together stay together. Is this an all-or-
none proposition? The Baldwins' (1969, 1973) squirrel monkey
data suggest that the correlation may not be too strong,
whereas Poirier's (1970) Nilgiri langur studies suggest it is.
If play functions in the socialization process, as well as in
social integration, then with the cessation of play, other
behaviors would be expected to increase or the social pattern
would be expected to change. The amount of play varies proportionately with the number and degree of social bonds based
on familiarity and affection and also with the length of time
an individual remains a member of the social group. The
amount of play varies inversely with the amount and degree of
aggression among group members.

E. Behavioral Flexibility

16. Are animals that play more flexible in their behavioral
responses to unexpected conditions? If play facilitates
behavioral flexibility, then groups that play more should be

expected to show more flexibility as part of their adaptation. Can the degree of flexibility in adult behaviors be related to the amount of time spent playing as a preadult? Does the degree of adult behavioral flexibility (learned behavior) vary in direct proportion with the length of maturation and the concomitant length and variety of play opportunities available to preadults?

F. Play and Exploration

17. Are play and exploration one and the same? Can we find some way to differentiate the two?

VII. CONCLUSION

There is no single adaptive function of play behavior; play may affect several aspects of an individual primate's development. Not only does play have a role in cognitive and physical development by providing variable stimuli to the nervous system and promoting muscular activity, it is also a means of socializing an individual to its group by providing opportunities for physical contact, developing relationships based on familiarity and affection, and learning its status in the dominance hierarchy. Play behavior is a mechanism for a youngster's adaptation to its physical and social environment, and, at the same time, it has adaptive value for the entire group by insuring social integration and providing variable behavior patterns for experimentation and innovation.

REFERENCES

Altmann, S. A. Sociobiology of rhesus monkeys. II: Stochastics of social communication. *J. theoret. Biol. 8,* 490-522 (1965).
Baldwin, J. D. The ontogeny of social behaviour of squirrel monkeys (*Saimiri sciureus*) in a semi-natural environment. *Folia primat. 11,* 35-79 (1969).
Baldwin, J. D., and Baldwin, J. I. The role of play in social organizations: comparative observations of squirrel monkeys (*Saimiri*). Paper delivered at American Association of Physical Anthropologists, Dallas (1973).
_____ The role of learning phenomena in the ontogeny of exploration and play, *in* "Primate Bio-social Development" (S. Chevalier-Skolnikoff, and F. E. Poirier, eds.), pp. 343-406. Garland Publishing Co., New York (1977).

Bekoff, M. The development of social interaction, play and
metacommunication in mammals: an ethological perspective.
Quart. Rev. Biol. 47, 412-434 (1972).
_____ Social play and play soliciting by infant canids.
Amer. Zool. 14, 323-340 (1974).
_____ Socialization in mammals with emphasis on nonprimates,
in "Primate Bio-social Development" (S. Chevalier-Skolni-
koff, and F. E. Poirier, eds.), pp. 603-636. Garland Pub-
lishing Co., New York (1977).
Berkson, G. The social ecology of defects in primates, *in*
"Primate Bio-social Development" (S. Chevalier-Skolnikoff,
and F. E. Poirier, eds.), pp. 189-204. Garland Publishing
Co., New York (1977).
Burton, F. D. The integration of biology and behavior in the
socialization of *Macaca sylvana* of Gibraltar, *in* "Primate
Socialization" (F. E. Poirier, ed.), pp. 29-62. Random
House, New York (1972).
Burton, F. D., and Bick, M. J. A. A drift in time can define
a deme: the implication of tradition drift in primate
societies for hominid evolution. *J. hum. Evol. 1,* 53-59
(1971).
Campbell, B. G. "Human Evolution: An Introduction to Man's
Adaptations". Aldine, Chicago (1966).
Carpenter, C. R. A field study of the behavior and social re-
lations of howling monkeys. *Comp. Psychol. Monogr. 10,*
1-168 (1934).
Chalmers, N. Comparative aspects of early infant development
in captive cercopithecines, *in* "Primate Socialization"
(F. E. Poirier, ed.), pp. 63-82. Random House, New York
(1972).
Chevalier-Skolnikoff, S. Visual and tactile communication in
Macaca arctoides and its ontogenetic development. *Am. J.
phys. Anthrop. 38,* 515-518 (1973).
_____ A Piagetian model for describing and comparing sociali-
zation in monkey, ape and human infants, *in* "Primate Bio-
social Development" (S. Chevalier-Skolnikoff, and F. E.
Poirier, eds.), pp. 159-187. Garland Publishing Co., New
York (1977).
Chevalier-Skolnikoff, S., and Poirier, F. E. (eds.), "Primate
Bio-social Development". Garland Publishing Co., New York
(1977).
Chomsky, N. "Aspects of the Theory of Language". Massachu-
setts Institute of Technology Press, Boston (1965).
Crook, J. H. The socio-ecology of primates, *in* "Social Behav-
ior of Birds and Mammals" (J. H. Crook, ed.), pp. 103-166.
Academic Press, New York (1970).
DeVore, I., and Eimerl, S. "The Primates". Time-Life Books,
New York (1965).

Diamond, E. "The Social Behavior of Mammals". Harper, New York (1970).

Diamond, S., and Hall, W. Evolution of the neocortex. *Science* *164*, 251-262 (1969).

Dolhinow, P. J., and Bishop, N. The development of motor skills and social relationships among primates through play. *Minn. Symp. on Child Psychol.* *4*, 141-198 (1970).

Etkin, W. (ed.), "Social Behavior from Fish to Man". University of Chicago Press, Chicago (1967).

Fedigan, L. N. Social and solitary play in a colony of vervet monkeys *(Cercopithecus aethiops)*. *Primates* *13*, 347-364 (1972).

Fedigan, L. N., and Fedigan, L. The social development of a handicapped infant in a free-living troop of Japanese monkeys, *in* "Primate Bio-social Development" (S. Chevalier-Skolnikoff, and F. E. Poirier, eds.), pp. 205-222. Garland Publishing Co., New York (1977).

Frisch, J. E. Individual behavior and intertroop variability in Japanese macaques, *in* "Primates: Studies in Adaptation and Variability" (P. Jay, ed.), pp. 243-252. Holt, Rinehart and Winston, New York (1968).

Gibson, K. Brain structure and intelligence in macaques and human infants from a Piagetian perspective, *in* "Primate Bio-social Development" (S. Chevalier-Skolnikoff, and F.E. Poirier, eds.), pp. 113-157. Garland Publishing Co., New York (1977).

Goodall, J. Chimpanzees of the Gombe Stream Reserve, *in* "Primate Behavior: Field Studies of Monkeys and Apes" (I. DeVore, ed.), pp. 425-473. Holt, Rinehart and Winston, New York (1965).

Goy, R. W., and Phoenix, C. The effects of testosterone propionate administered before birth on the development of behavior in genetic female rhesus monkeys, *in* "Steroid Hormones and Brain Function" (C. Sawyer, and R. Gorski, eds.), pp. 193-200. University of California Press, Berkeley (1972).

Goy, R. W., and Resko, J. Gonadal hormones and behavior of normal and pseudohermaphroditic nonhuman female primates, *in* "Recent Progress in Hormone Research", Vol. 28 (E. B. Astwood, ed.), pp. 707-733. Academic Press, New York (1972).

Goyer, R. Communication, communicative process, meaning: toward a unified theory. *J. Communication* *20*, 4-16 (1970).

Hall, K. R. L. The behaviour and ecology of the wild patas monkey, *Erythrocebus patas*, in Uganda. *J. Zool.* *148*, 15-87 (1965).

Hall, K. R. L., and Goswell, M. Aspects of social learning in captive patas monkeys. *Primates* *5*, 59-70 (1964).

Harlow, H. F.   Basic social capacity of primates, *in* "Primate
    Social Behavior" (C. H. Southwick, ed.), pp. 153-160. D.
    Van Nostrand Co., Princeton (1963).
_____   Age-mate or peer affectional systems, *in* "Advances in
    the Study of Behavior", Vol. 2 (D. S. Lehrman, R. A.
    Hinde, and E. Shaw, eds.), pp. 334-384. Academic Press,
    New York (1969).
Harlow, H. F., and Harlow, M. K.   A study of animal affection.
    Nat. Hist. 70, 48-55 (1961).
_____   Learning to love. *Amer. Sci. 54*, 244-272 (1966).
Harlow, H. F., Harlow, M. K., and Suomi, S.   From thought to
    therapy: lessons from a primate laboratory. *Amer. Sci.
    59*, 538-549 (1971).
Harman, P.   Paleoneurologic, neoneurologic and ontogenetic
    aspects of brain phylogeny.   James Arthur lecture on the
    evolution of the human brain. American Museum of Natural
    History, New York (1957).
Hinde, R. A., and Spencer-Booth, Y.   The behaviour of socially
    living rhesus monkeys in their first two and a half years.
    Anim. Behav. 15, 169-196 (1967).
Itani, J.   On the acquisition and propagation of a new food
    habit in the natural groups of the wild Japanese monkey at
    Takasakiyama. *Primates 1-2*, 84-98 (1958).
Itani, J., and Nishimura, A.   The study of infrahuman culture
    in Japan: a review, *in* "Precultural Primate Behavior",
    Symp. IVth Int. Congr. Primat., Vol. 1 (E. Menzel, ed.),
    pp. 26-50. Karger, Basel (1973).
Jay, P. (ed.), "Primates: Studies in Adaptation and Variabil-
    ity". Holt, Rinehart and Winston, New York (1968).
Jolly, A.   "The Evolution of Primate Behavior". Macmillan,
    New York (1972).
Kalmus, H.   Animal behaviour and theories of games and of
    language. *Anim. Behav. 17*, 607-617 (1969).
Kawai, M.   Newly acquired precultural behavior of the natural
    troop of Japanese monkeys on Koshima Islet. *Primates 6*,
    1-30 (1965).
Knudson, M.   Sex differences in dominance behavior of young
    human primates. Paper delivered at the American Anthro-
    pological Association, New York (1971).
_____   Sex differences in dominance behavior of young human
    primates. Unpublished Ph.D. Dissertation, University of
    Oregon, Eugene (1973).
Kummer, H.   Two variations in the social organization of
    baboons, *in* "Primates: Studies in Adaptation and Varia-
    bility" (P. Jay, ed.), pp. 293-312. Holt, Rinehart and
    Winston, New York (1968).
_____   "Primate Societies". Aldine Publishing Co., Chicago
    (1971).

Lancaster, J. Play-mothering: the relations between juvenile
    females and young infants among free-ranging vervet mon-
    keys (Cercopithecus aethiops). Folia primat. 15, 161-183
    (1971).
_____ Play-mothering: the relations between juvenile females
    and young infants among free-ranging vervet monkeys, in
    "Primate Socialization" (F. E. Poirier, ed.), pp. 83-104.
    Random House, New York (1972).
Lindburg, D. G. The rhesus monkey in North India: an ecologi-
    cal and behavioral study, in "Primate Behavior: Develop-
    ments in Field and Laboratory Research", Vol. 2 (L. A.
    Rosenblum, ed.), pp. 1-106. Academic Press, New York
    (1971).
Loy, J., and Loy, K. Behavior of an all juvenile group of
    rhesus monkeys. Am. J. phys. Anthrop. 40, 83-96 (1974).
Mason, W. A. The effects of environmental restriction on the
    social development of rhesus monkeys, in "Primate Social
    Behavior" (C. H. Southwick, ed.), pp. 161-173. D. Van
    Nostrand, Princeton (1963).
_____ The social development of monkeys and apes, in "Primate
    Behavior: Field Studies of Monkeys and Apes" (I. DeVore,
    ed.), pp. 514-543. Holt, Rinehart and Winston, New York
    (1965).
_____ Naturalistic and experimental investigations of the
    social behavior of monkeys and apes, in "Primates: Studies
    on Adaptation and Variability" (P. Jay, ed.), pp. 398-419.
    Holt, Rinehart and Winston, New York (1968).
McGrew, W. Socialization and object manipulation of wild
    chimpanzees, in "Primate Bio-social Development" (S. Chev-
    alier-Skolnikoff, and F. E. Poirier, eds.), pp. 261-288.
    Garland Publishing Co., New York (1977).
Menzel, E. Responsiveness to objects in free-ranging Japanese
    monkeys. Behaviour 26, 130-150 (1966).
_____ Chimpanzee utilization of space and responsiveness to
    objects: age differences and comparison with macaques,
    in "Proc. of 2nd Int. Congr. Primat.", Vol. I (C. R.
    Carpenter, ed.), pp. 72-80. Karger, Basel (1969).
Norback, E., and Moskowitz, N. The primate nervous system:
    functional and structural aspects of phylogeny, in "Evo-
    lutionary and Genetic Biology of Primates", Vol. 1 (J.
    Buettner-Janusch, ed.), pp. 131-175. Academic Press, New
    York (1963).
Parker, S. Piaget's sensory motor series in an infant
    macaque. A model for comparing unstereotyped behavior and
    intelligence in human and nonhuman primates, in "Primate
    Bio-social Development" (S. Chevalier-Skolnikoff, and F. E.
    Poirier, eds.), pp. 43-112. Garland Publishing Co., New
    York (1977).

Phoenix, C., Goy, R., and Resko, J. Psychosexual differenta-
    tion as a function of androgenic stimulation, *in* "Perspec-
    tives in Reproduction and Sexual Behavior" (M. Diamond,
    ed.), pp. 33-49. Indiana University Press, Bloomington
    (1968).
Poirier, F. E. Analysis of a Nilgiri langur (*Presbytis
    johnii*) home range change. *Primates 9*, 29-44 (1968a).
_____ The Nilgiri langur (*Presbytis johnii*) mother-infant
    dyad. *Primates 9*, 45-68 (1968b).
_____ The Nilgiri langur troop: its composition, structure,
    function and change. *Folia primat. 11*, 20-47 (1969a).
_____ Nilgiri langur (*Presbytis johnii*) territorial behavior.
    Primates 9, 351-365 (1969b).
_____ Behavioral flexibility and intertroop variability among
    Nilgiri langurs of South India. *Folia primat. 11*, 119-133
    (1969c).
_____ Nilgiri langur ecology and social behavior, *in* "Primate
    Behavior: Recent Developments in Field and Laboratory Re-
    search", Vol. I (L. A. Rosenblum, ed.), pp. 251-383.
    Academic Press, New York (1970).
_____ Socialization variables. Paper read at the American
    Anthropological Association, New York (1971).
_____ Introduction, *in* "Primate Socialization" (F. E. Poirier,
    ed.), pp. 2-29. Random House, New York (1972a).
_____ Nilgiri langur behavior and social organization, *in*
    "Essays to the Chief" (F. Voget, and R. Stephenson, eds.),
    pp. 119-134. University of Oregon Press, Eugene (1972b).
_____ Primate socialization and learning. Paper read at the
    American Anthropological Association, Toronto (1972c).
_____ Primate socialization - where do we go from here?
    Paper read at the American Association of Physical Anthro-
    pologists, Dallas (1973a).
_____ Primate socialization and learning, *in* "Culture and
    Learning" (S. Kimball, and J. Burnett, eds.), pp. 3-41.
    University of Washington Press, Seattle (1973b).
_____ Colobine aggression: a review, *in* "Primate Aggression,
    Territoriality, and Xenophobia: A Comparative Perspective"
    (R. L. Holloway, ed.), pp. 123-158. Academic Press, New
    York (1974).
_____ Introduction, *in* "Primate Bio-social Development" (S.
    Chevalier-Skolnikoff, and F. E. Poirier, eds.), pp. 1-39.
    Garland Publishing Co., New York (1977).
Poirier, F. E., and Smith, E. O. Socializing functions of
    primate play behavior. *Amer. Zool. 14*, 275-287 (1974).
Radinsky, L. B. Primate brain evolution. *Amer. Sci. 63*, 656-
    663 (1975).
Ransom, T., and Rowell, T. E. Early social development of
    feral baboons, *in* "Primate Socialization" (F. E. Poirier,
    ed.), pp. 105-144. Random House, New York (1972).

Rosenblum, L. A. The ontogeny of mother-infant relations in macaques, in "Ontogeny of Vertebrate Behavior" (H. Moltz, ed.), pp. 315-367. Academic Press, New York (1971).

Rowell, T. E. Forest-living baboons in Uganda. J. Zool. Lond. 149, 344-364 (1966).

Rumbaugh, D. M. Learning skills of anthropoids, in "Primate Behavior: Recent Developments in Field and Laboratory Research", Vol. 1 (L. A. Rosenblum, ed.), pp. 1-70. Academic Press, New York (1970).

Schultz, A. Postembryonic age changes, in "Primatologia" (H. Hofer, A. Schultz, and D. Starck, eds.), pp. 887-964. S. Karger, Basel (1956).

Simonds, P. "The Social Primates". Harper and Row, Boston (1974).

Smith, E. O. The interrelationship of age and status and selected behavioral categories in male pigtail macaques (Macaca nemestrina). Unpublished Masters Thesis. University of Georgia, Athens (1972).

_____ A model for the study of social play in non-human primates. Paper read at Midwest Animal Behavior Society, Oxford, Ohio (1973).

_____ Social play in rhesus macaques (Macaca mulatta). Unpublished Ph.D. Dissertation. Ohio State University, Columbus (1977).

Sorenson, M. Behavior of tree shrews, in "Primate Behavior: Developments in Field and Laboratory Research", Vol. 1 (L. A. Rosenblum, ed.), pp. 141-192. Academic Press, New York (1970).

Struhsaker, T. Behavior of vervet monkeys, Cercopithecus aethiops. Univ. Calif. Publ. Zool. 82, 1-74 (1967).

Sussman, R. Socialization, social structure, and ecology of two sympatric species of Lemur, in "Primate Bio-social Development" (S. Chevalier-Skolnikoff, and F. E. Poirier, eds.), pp. 515-528. Garland Publishing Co., New York (1977).

Symons, D. Aggressive play in a free-ranging group of rhesus monkeys (Macaca mulatta). Unpublished Ph.D. Dissertation, University of California, Berkeley (1973).

Tisza, V., Hurwitz, I., and Angoff, K. The use of a play program by hospitalized children. J. Am. Acad. Child Psychiat. 9, 515-531 (1970).

Tsumori, A. Newly acquired behavior and social interactions of Japanese monkeys, in "Social Communication Among Primates" (S. A. Altmann, ed.), pp. 207-219. University of Chicago Press, Chicago (1967).

Tsumori, A., Kawai, M., and Motoyishi, R. Delayed response of wild Japanese monkeys by the sand-digging test (I) -- case of the Koshima troop. Primates 6, 195-212 (1965).

van Lawick-Goodall, J. Mother-offspring relationships in free-ranging chimpanzees, *in* "Primate Ethology" (D. Morris, ed.), pp. 287-347. Aldine, Chicago (1967).

Washburn, S. L. Primate field studies and social science, *in* "Cultural Illness and Health. Essays in Adaptation" (L. Nader, and T. Maretzki, eds.), pp. 128-134. American Anthropological Association, Washington (1973).

Washburn, S., and DeVore, I. The social life of baboons. *Sci. Amer.* 204, 62-71 (1961).

Washburn, S., and Hamburg, D. The implications of primate research, *in* "Primate Behavior: Field Studies of Monkeys and Apes" (I. DeVore, ed.), pp. 607-622. Holt, Rinehart and Winston, New York (1965).

Welker L. Ontogeny of play and exploratory behavior: a definition of problems and research for new conceptual solutions, *in* "The Ontogeny of Vertebrate Behavior" (H. Moltz, ed.), pp. 171-228. Academic Press, New York (1971).

White, R. Motivation reconsidered: the concept of competence. *Psych. Rev.* 66, 297-333 (1959).

THE FUNCTION OF ADULT PLAY IN
FREE-RANGING *Macaca mulatta*

J. A. Breuggeman

Department of Sociology and Anthropology
Purdue University
West Lafayette, Indiana

Play behavior is perhaps most reasonably viewed as a multifunctional and multicausal behavior not only at the interspecific level, but at the intraspecific level as well, especially among complex, socially living species. Among primates, the function of play is usually described as inducing and/or facilitating 'learning' or 'socialization' in the immature individuals.

Observation of a free-ranging social group of rhesus monkeys (Macaca mulatta) over a 15-month period indicates that the occurrence of play in adult rhesus, although rare, is common enough to necessitate at least a modification of this largely assumed function of play behavior. These observations offer support for the idea that play among rhesus monkeys is a multifunctional behavior. Contextual observations of playing adult rhesus and the animals they played with suggest that social manipulation is a potentially important function of play at all stages of the life cycle and particularly among adults.

I. INTRODUCTION

In the past, there have been many suggestions that the function of play behavior is not the same for all species. More recently, suggestions have been made concerning possible variation in play function within a species (cf., Aldis, 1975). An examination of data derived from a fifteen-month study of a social group of free-ranging rhesus monkeys on Cayo Santiago presents the likelihood of such variability in play function.

The focus of this study will concern the play involving
adult animals; animals who presumably have no need to train or
exercise, whose physical development is complete, who already
know their environment, their social skills, and their rank
order relationships, and whose social bonds are somewhat predictable, even in those few cases where they are not well
established. Adult play in the study group, although rare,
was by no means nonexistent. Further, it may be determined by
easily identifiable characteristics, such as form, kind and
manner of occurrence. The probable function of such play is
best suggested by the contexts in which it occurs.

II. THE QUESTION OF ADULT PLAY

Most play is indeed among young animals. The fact that
this is not the only occurrence of play has until recently been
rarely mentioned (Breuggeman, 1976). Aldis (1975) devoted a
brief chapter in his recent book to the question of adult play.
He feels that among primates, the most important form for consideration is that of mother-infant play, and this only among
those primates who live in nonstable social groups or small
social units, i.e., chimpanzees and the prosimians. These are
not the natural social situations of the rhesus monkey.
Generally, given our current theoretical framework, there
is little or no reason why adult primates should play. In
fact, Aldis has dismissed most of the reports of this behavior
by pointing out that adult males have been reported to play
more than adult females and since there are probably discrepan-
cies in the age a male is considered as worthy of the label
'adult', such reports are unreliable or, to use his words, "to
be taken with a grain of salt" (Aldis, 1975:110). Discrepan-
cies of former studies aside, the following study attempts to
describe the changes in play behavior in later ontogeny by
focusing on the context in which this behavior occurs, and by
giving some attention to the possible meaning of individual
bouts of this behavior.
Usage of the term 'context' began in linguistics in the
1930's. For over a decade now, it has been an important concept for the ethologist also. It has been used primarily for
the analysis of animal communication systems. In this sense,
W. John Smith, in 1965, stated: "Since in all natural situations, both immediate and historical contexts of signals are
always present, the term 'context' can be used to refer to
their combined influences. This ordinary sense of 'context'
refers to anything which can be thought of as accompanying a
signal" (p. 405). It has been demonstrated that even in rela-

tively simple vertebrate communication systems, the meaning of
a signal will vary according to the context in which it appears.
Bramblett (1976) has pointed out that modern field studies
in primatology have intensified the realization that all behavior occurs in a context and that due to the complexity of primate behavioral systems and the rapidity with which behavioral
events occur, the meaning of the behavioral events themselves
should not be left solely to the intuition of the observer.

Although it is undoubtedly important to be aware of the effects of the environmental context, this is frequently a difficult concept to deal with. Awareness of social and individual
contexts is somewhat easier, and factors such as sex, rank and
age lend themselves more easily to analysis. Frequently, these
and similar delineations are not adequate to promote an awareness of what a particular bout of behavior might mean.

The immediate biological context of an organism may possibly also contribute to the occurrence of a particular behavior, but under field conditions, we must rely largely on seasonal and/or maturational indicators to generate our hypotheses.

All of these contexts should be taken into consideration
in the analysis of behavior. It is likely that in some cases,
we should take this idea of context one step further. That is,
in viewing bouts of certain behaviors, we need to not only have
a knowledge of the overall social system, as well as knowledge
of the individuals involved in the behavior and an adequate
description of the bout of behavior in question, but we need
also to consider the bouts of behavior that immediately precede
and follow the bout.

This kind of sequential perspective is not particularly innovative, but it is one that has been frequently ignored in the
analysis of the behavior of complex social organisms. It is a
perspective that should, however, contribute to our understanding of what individual bouts of behavior might mean to the organisms involved.

The question of meaning has frequently been unfashionable
in ethological studies, and for good reason, since we have a
history laden with anthropomorphic anecdotes, especially in
primatology. Given our advance in methodology and theory, it
would seem reasonable at this time to attempt some answer to
this question, especially in those areas of behavior study
where our past researches have been largely unfruitful in ad-
dressing basics such as function, causation, and survival
value. The area of play research is one such area.

Huizinga (1950) said, "All play means something". Good
description based on adequate observation is basic to any ethological endeavor, but good ethology should be more than good
description. There is fair agreement now that play may be
identified in many species by rotational movement in the trans-

verse plane (Aldis, 1975). On what this movement means when
it occurs, there is no agreement.

Many have suggested that play behavior is a multifunctional
and possibly a multicausal behavior. Possible substantiation
for this perspective has been rare, especially at the intraspecific level.

III. DESCRIPTION OF STUDY

In June of 1969, I began my study of group F, a free-ranging group of *Macaca mulatta* of the Cayo Santiago, Puerto Rico,
colony. Birth records have been kept on the group since 1956
and, hence, the genealogical relationships of all but the oldest females are known, as well as the age of each individual
born after 1956 in group F (see Figure 1). All individuals
were identifiable by tattoos on their chests and medial thighs.
A system of ear-notching also aided in identification, until I
was able to recognize the individuals on sight.

Note-taking techniques consisted of collecting all episodes of dominance, grooming, copulation, mounting, body contact and play. Beyond that, I attempted to record all that I
saw in long-hand notes taken at the time of observation. Attempts were made to observe all group members equally during
each day's observations. Of special interest were the contexts
of behavioral episodes and whenever possible, these data were
recorded. During the study period on Cayo Santiago, group F
was observed over a continuous 15-month period for a total of
1,600 hours of direct observation.

A. *Composition of Sample*

The sample for the study of adult play includes only those
animals that were older juveniles or adults at the time of the
onset of the study; that is, all animals three years or older
as of June, 1969 (see Figures 2 and 3). This was done for two
reasons: 1) it would best show the contrast between young
adults, mature adults and old adults; and 2) in terms of reproductive behavior (cf., Loy, 1969), all of the three and four
year olds were sexually active during the 1969 mating season
and it, therefore, seemed reasonable to include them in the
study.

These animals were involved in a total of 2,140 bouts of
play. This represents about 25% of the total play seen in the
study group during the fifteen-month period. All of these animals, with the exception of two low-ranking old adult females,
were observed to play at some time during the study period.

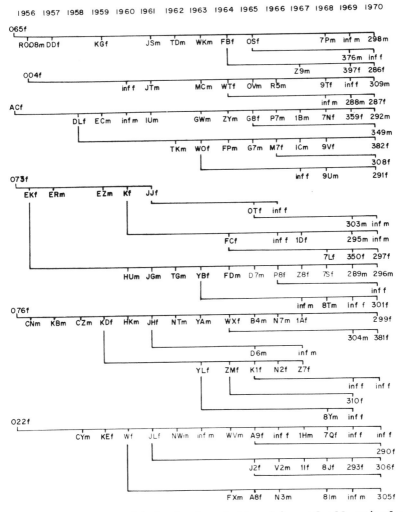

FIGURE 1. *Known biological relationships of all animals born in group F since 1956. Genealogies are arranged according to rank, highest to lowest, top to bottom.*

IV.   ADULT RHESUS PLAY: DESCRIPTION

    Play involving adult rhesus shows a number of interesting
characteristics. Like the play of the young, adult play may
be generally characterized by an emphasis on rotary movements
in the transverse plane (cf., Aldis, 1975; Sade, 1973); however, it is not nearly as varied in form, kind and, presumably,
function as the play of the young. Adults were never observed
to be involved in either solitary play or object play. Further, they were never seen involved in complex social play
(i.e., "games"), but only simple kinds of social play: wrestling and chasing. Both body movement and the movement of the
body through space were very limited. In fact, the observed
play bouts of adults were usually stationary. Further, especially in the case of fully mature adults, the animals involved
in such bouts nearly always appeared to have a very low state
of arousal, and the movements used usually had a 'slow motion'
quality to them. These characteristics in combination gave
the play activities of adults a very 'deliberate' or 'premedi-
tated' appearance.

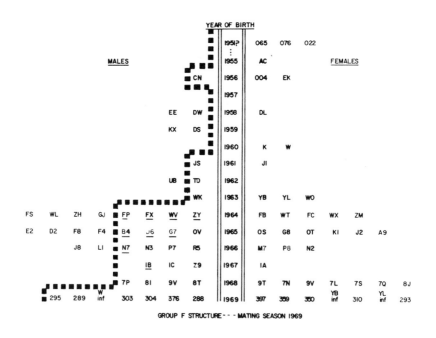

*FIGURE 2. Underlined names are those of animals that died
or disappeared from group F during the period of study. Males
included within the dotted line are all natal males.*

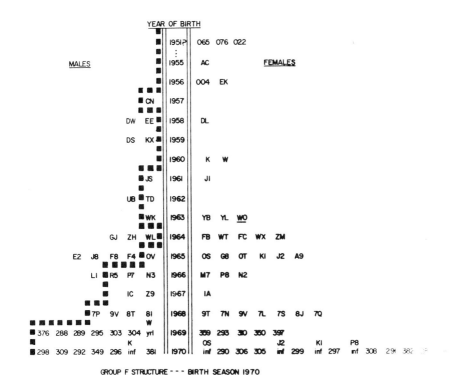

GROUP F STRUCTURE - - - BIRTH SEASON 1970

FIGURE 3. *Underlined names are those of animals that died
or disappeared from group F during the period of study. Males
included within the dotted line are all natal males.*

Adult play was nearly always seen to occur on the ground,
rarely in the water (and then, only among young adults), and
never in the trees. As previously mentioned, only simple kinds
of social play were observed among adults and among fully mature adults only very rarely did these bouts occur in the context of a play group, or in the presence of other playing animals. Once past young adulthood, any play was extremely rare.

V. EFFECT OF SELECTED VARIABLES

The effects of the variables of age, sex and genealogy on
the appearance of adult play give many clues to the nature of
rhesus play and its importance in the life cycle of the species.

A. Age

The effect of age on adult play is quite dramatic. As mentioned, the number of play bouts which involved adults accounted for only about 25% of the total play observed during the
period of study. This is somewhat misleading, however, if a
breakdown of the age groups is not considered. For example,
although three-year-olds were only considered 'adults' during
the period of the mating season, the three-year-olds accounted
for over 30% of the total adult play seen. Four-year-olds for
the entire period of study accounted for another 35% of the
total adult play seen (see Figure 4). In other words, the
three-year-olds during the mating season and the four-year-olds
during the entire period of study account for about 16% of the
total play observed. The five- through eight-year-olds account
for almost another 30% of the total adult play seen, or less
than 8% of the total play. This means that the animals aged
nine years and older account for only a little over 5% of the
adult play (see Figure 4).

Three- and four-year-olds accounted for almost 85% of the
adult play seen during the mating season. Basically, this
same group of animals as four- and five-year-olds in the birth

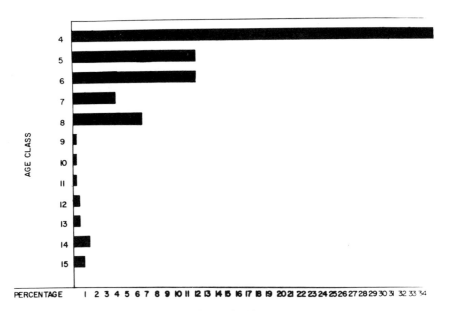

PERCENTAGE PER ADULT AGE CLASS OF TOTAL ADULT PLAY

*FIGURE 4. All adult age classes were seen to play during
the study period.*

season accounted for 70% of all adult play. The five-, six- and seven-year-olds in the mating season accounted for 11% of the adult play but in the birth season as six-, seven- and eight-year-olds, basically this same group of animals accounted for 24% of this play. Animals eight years of age and over accounted for about 3% of adult play in the mating season, and those nine years of age and over about 5% of this play during the birth season. In other words, the distributions do seem to distinguish the categories 'young adult', 'mature adult', and 'old adult', and it does appear that mature adults and old adults are seen to play slightly more often in the birth season, although the skewed nature of the raw frequencies seems to indicate that the effect of individual variation is fairly strong (see Figures 5 and 6).

The most interesting effect of age is seen in consideration of who the adults play with. During the mating season, 75% of all play bouts involving adults had an animal under three years of age as the other member of the dyad. In other words, only 25% of the play involved other adults. During the birth season, 73% of play bouts involving adults were with animals two years of age or younger.

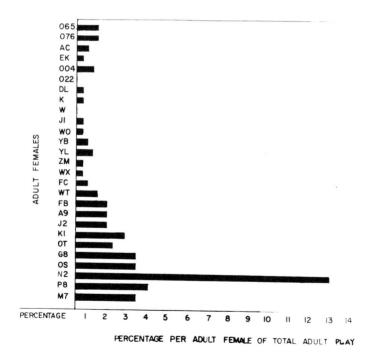

FIGURE 5. Only two adult females of group F were not seen to be participants in play bouts during the period of study.

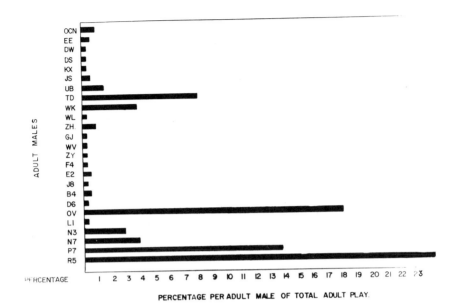

FIGURE 6. All adult males, both natal and non-natal, were seen to play during the period of study.

During the period of study, there were 500 bouts of play involving adults where the initiator of the bout was known. Three hundred eleven (62%) of these bouts were initiated by the older animal of each dyad. Of these bouts initiated by older animals, 63% were directed toward animals two years of age or younger. Only nine bouts of play were observed to be initiated by animals nine years of age or over with animals three years of age or older. In other words, older adults basically do not initiate bouts of play with younger adults. When this occurs, the context of the interaction is of special interest and, as such, will be discussed later.

Seventy-six percent of all bouts where the initiator was known were initiated by three-, four-, or five-year-olds. These bouts initiated by young adults, interestingly enough, were initiated with animals of all ages, from the very youngest to the very oldest.

It appears, therefore, that the effect of age on adult play is considerable. To be fully understood, it must be considered in conjunction with other variables.

B. *Sex and Genealogical Relationships*

One of the most frequently reported differences in
male and female primate young occurs in play activities.
Males and females are reported to exhibit differences in the
frequency with which play occurs and the kind of play that occurs. As previously mentioned, the kind of play that is observed among mature adults and old adults is usually the same:
low intensity, stationary, noncomplex social play. In other
words, it is similar not only in kind but also in form.
The frequencies of play observed among all the animals that
were considered adults did, however, show differences. Considered as a class, the males were observed to play about twice
as much as the females. When these percentages were considered
in terms of the number of adult females and the number of adult
males in the group, the relative amount of play shown for each
individual male was somewhat less than twice the amount for
each individual female.
Adults played with males about 70% of the time during both
the mating and the birth season. Adult males did play with
both male and female animals, but mature adult and old adult
males were never seen to play with mature adult females. Young
adult females were seen on rare occasions to play with young
adult males. Young adult females, as with young adult males,
were more likely to be seen playing with animals of any age/sex
class. Some young adult female individuals were observed to
play more than some young adult males. This is especially
true when comparing young adult females with non-natal males
of the same age and, in some cases, it is true in comparison
with natal males.
Sex differences in the play behavior of adult rhesus would,
therefore, appear to be minimal, especially among the old
adults, where both males and females were observed to play
with about the same frequency.
Genealogical relationships also affect adult play. There
seems to be a slight tendency for higher ranking adult animals
to play more than lower ranking adult animals and, certainly,
for natal adults to play more than non-natal adults. Further,
the adjusted percentages per number of individuals in each
genealogy do not change this pattern greatly (see Figure 7).
These patterns were essentially the same for both the birth
season and the mating season.
Genealogical relationships do appear to affect the appearance of adult play (see Figure 8). Clearly, high ranking animals are involved in play more often than animals of low ranking genealogies. This is somewhat misleading, however. More
accurately, adults of high ranking genealogies (see Figure 1)
play more often and their play involves animals of their own
genealogy.

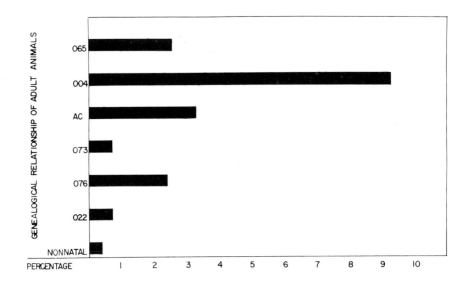

RELATIVE AMOUNT OF PLAY OBSERVED PER INDIVIDUAL
FOR EACH GENEALOGY. THIS IS EXPRESSED IN TERMS
OF THE AVERAGE PERCENTAGE PER INDIVIDUAL IN EACH GENEALOGY

*FIGURE 7. Adult individuals of higher ranking genealogies
tended to play more than individuals of lower ranking genealogies. Non-natal male individuals, considered as a class,
played less frequently than the adults of the lowest ranking
genealogy.*

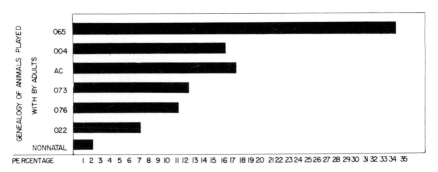

PERCENTAGE PER GENEALOGY OF ANIMALS PLAYED WITH BY ADULTS

*FIGURE 8. Adult animals of higher ranking genealogies not
only played more frequently, but played more often with animals of their own genealogy.*

Of the twenty-five females in the group with offspring, seventeen of these females were seen to play with one, usually the youngest, of their offspring. These females ranged from high to low ranking, from the very oldest to the very youngest of mothers. All of the five females of the first and second ranking genealogies were observed to play with their young, as were three of the five mothers from the third ranking genealogy, four of the six mothers of the fourth ranking genealogy, four of the five mothers of the fifth ranking genealogy, and one of the five mothers of the sixth ranking genealogy. As previously mentioned, two of these females (022 and W) from this lowest ranking genealogy were never seen to play. Another four of the eight females who were not observed to play with their offspring were the females who gave birth for the first time during the latter portion of the study. In one of these cases, the infant only lived for four months; in another, the infant was born very late in the season.

Mother/offspring play accounted for only a small percentage of the total adult play seen, however, with only 81 bouts observed for the entire period of study. There was no apparent seasonal difference in the frequency of mother/offspring play.

With the exception of females 065 and JI, all females that were seen to play with their offspring were only observed playing with the very youngest of their offspring, that is, infants or yearlings. For the majority of these dyads, less than five bouts of play were observed.

One of the dyads, ZM and her yearling 310, to be discussed later, accounted for 30% of the observed mother/offspring play bouts. During the mating season, these bouts between ZM and 310 accounted for only 8% of the total mother/offspring bouts; during the birth season, they accounted for 50% of the total number of these bouts. It was during this time that ZM, who did not give birth that spring, was trying rather unsuccessfully to wean the yearling 310.

Play with younger siblings accounted for a much greater percentage of total adult play than did mother/offspring play, that is, 23% as compared to 4%. Overall seasonal differences are minimal here also. Of the 40 sibling dyads involving adult play, five dyads accounted for 50% of the observed bouts. All but one of these dyads involved young adults. The play between R5 and OV, both young adult natal males, who did not disperse from the group at puberty, accounted for 16% of the adult sibling play observed. Two-thirds of these bouts occurred during the mating season, presumably before the younger R5's rise in rank over OV had fully stabilized.

These patterns are merely suggestive, when the question of function is considered. A further consideration of the immed-

iate behavioral context surrounding the individual bouts of
play involving adults gives a clearer view of probable function.

VI. CONTEXT

   The social and/or behavioral context was recorded for only
5% of the play bouts in which adults were observed. While this
is a very small sample of an already small sample, still nearly
all of these contexts seem relatively important in that they
offer a variety of possible immediate functions of the play
involving adults.
   As previously mentioned, the form and kind of play exhibited by rhesus is essentially identical in all bouts. For example:
   7 August 1969 - *0922*: Old adult female 065 moves slowly
   to four-year-old daughter OS, places one arm around OS's
   neck and presses OS to the ground while rolling her head
   and mouthing OS slowly for a minute or so. 065 moves
   off from OS.
The play signals used by the young are also used by the adults:
   30 June 1969 - *0907*: Mature adult female K moves near
   male infant 288, turns, places her head and shoulder
   to the ground, looks back at 288 and then moves to and
   wrestles with 288.
Some bouts of adult play are extremely low key:
   2 July 1970 - *0726*: Female infant P870F gambols up to
   old adult female 076, who holds out her hand and rotates
   it rapidly. While P870F wrestles actively with 076's
   moving hand, the rest of 076's body remains still.
On some occasions, however, adults do wrestle actively. For
example, on 12 September 1969 - *0705*, the adult female FB was
seen wrestling actively with her younger brother 7P in several
bouts that were separated by episodes of leaping and rearing
and chasing. This is, however, rare for adults, especially
older adults.
   Other kinds of play, while they appear to be an attraction
for many adults, are not usually engaged in by adults:
   2 June 1970 - *0910*: A group of twelve yearlings, two-
   and three-year-old animals, gathered at a rain water
   pond, dive, swim and chase in the water. Mature adult
   male WK moves into the tree the animals are diving from
   and sits looking around. Young adult male OV moves to
   the area and sits on the edge of the pond watching the
   activity. WK climbs down from the tree and sits at the
   edge of the water batting a floating stick and then
   walks across the pond submerging himself more than

neceasary in the process and then sits at the other
side of the pond, while turning a rock over and over
again. OV moves to WK with tail whipping movements
and grinning, and bounces on his front legs several
times while looking at WK. OV then turns and looks at
WK from between his hind legs. WK looks at OV briefly
and then moves off. Adult female OT is also sitting at
the edge of the pond watching the play activities while
occasionally splashing her hand in the water. Young
adult N2 moves to OT and sits splashing her hand in
water also, at which point OT moves from the area.
This observation, and others like it, indicate that while there
are kinds of play which adults to not join, they still seem to
find these activities attractive and, at times, even tempting.

Within these parameters, the following specific contexts
may be delineated:

A. Mother/offspring and other parental dyads
B. 'Sibling rivalry' and other 'jealousies'
C. Sexual contexts
D. Agonistic contexts

A. Mother/Offspring and Other Parental Dyads

As previously mentioned, 4% of the observed adult play occurred between mothers and offspring, usually the youngest of
the offspring. One of the most obvious contexts for this kind
of play dyad is the weaning process. The weaning process is
extremely gradual among rhesus monkeys and begins very early
in the infant's life (Breuggeman, 1973). Play behavior, apparently functioning as a distraction tactic, has been seen to
occur very early in this context. For example, on 6 July 1970
- *1634,* when infant 349 was less than two months old, he was
observed to run to his mother, G8, and grab for her nipple,
but instead of the adult female allowing the infant to suckle,
she grabbed him, pushed him away and wrestled with him. This
female, however, was a young mother. It should be noted that
in all of the mother/offspring play bouts that occurred during
the weaning process, only young mothers were involved. Further, in about half of the dyads where play was seen in this
context, the offspring of the dyad was a yearling. For example, on 24 July 1969 - *0720,* shortly after the death of
YB69F (YB69F = YB's infant), yearling 8T reached for his
mother's nipple while she (YB) was grooming him. YB quickly
pushed 8T from her and began to wrestle with him vigorously
and then batted him away.

A high percentage of mother/offspring bouts occurred in
one mother/young dyad, ZM and 310, and are, therefore, of special interest. All play behavior that was recorded for this

dyad was noted in the context of the infant's attempts to
suckle. Interestingly enough, the play bouts that occurred in
this context showed a considerable range of intensity on the
part of the adult female. The play recorded earlier in the
study period was low arousal and 'typical' of play involving
an adult animal. Toward the end of the study, ZM was not only
playing very actively with 310 when the infant tried to suckle,
she appeared almost frantic. The following observation exem-
plifies the change:

2 July 1970 - *0800*: 310 grabs at ZM's nipple, ZM grabs
and wrestles 310 vigorously and then chases off. 310
chases to ZM, leaps and tags ZM, who turns, rears,
wrestles and then leaps from 310 and chases off. This
sequence is repeated several times as I watch the two
of them move into the distance, chasing, tagging, leaping and wrestling.

It is, of course, unusual to see an adult involved in play of
this duration and intensity. It should be noted that 310 was
seen to initiate play in this context with ZM as often as ZM
initiated the play bout.

Other young were also seen to initiate play with their
mothers when the nipple was refused:

2 July 1969 - *0744*: Infant 304 runs to WX and takes
her nipple in his mouth. WX quickly pushes 304 away.
304 leaps and wrestles WX, who begins to roll her head
and mouths at 304 briefly, then once again pushes 304
away and this time moves off.

Wrestling was also seen when an infant attempted to suckle
a female other than the mother:

9 April 1970 - *0822*: Infant 309, son of 004, moves to
and takes DL's nipple in his mouth as DL sits grooming
065. DL wrestles 309 until 309 moves off, moves to
and takes YB's nipple in his mouth just as 004 moves
into the area. 004 threatens YB with opened mouth as
YB cowers from 004 and moves off.

Many of the other observed contexts of adult play involved
essentially parental behaviors (cf., Breuggeman, 1973), although the adult involved was not the mother, but perhaps a
sibling or other group member.

Play was seen to occur most frequently in the context of
parental care when an adult male or female other than the mother tried to cuddle, collect or carry an infant that did not
appear to be interested in being cuddled, collected or carried.
The adult caretaker was frequently seen to resort to play behavior to restrain the infant. Some of the recorded contexts
were with siblings, but most were not. Frequently, the adult
caretaker was not genealogically related to the infant:

28 June 1970 - *1512*: OS collects and cuddles her infant brother 298, who immediately moves to and grooms

their mother 065. OS again moves to and cuddles 298.
298 begins to push at OS trying to get away, at which
point OS wrestles 298.

Infants were seen to initiate play with adults apparently
to avoid such caretaking behavior. This was seen to occur most
frequently in mother/offspring dyads:

25 April 1970 - *1032*: M7 grooms 308 as 308 tries to
scramble off. M7 grabs 308's leg, rolls her onto her
back and begins to groom her once more. 308 wrestles
M7's hands with both hands and feet as M7 continues to
try to groom 308.

This behavior is apparently not 'learned', or, if it is, it
is learned before other kinds of social play. Infant 349 was
observed to initiate such play with his mother twice during
the third week after his birth.

Play is seen not only between mothers and infant offspring,
but also with older offspring, at times much older than might
be expected, for example, with young adult offspring. Play
bouts with young adult offspring were always observed to be
initiated by old adult mothers.

30 October 1969 - *1035*: 004 grooms her young adult
son, R5, who, in turn, reaches out to his yearling
sister 9T and begins to wrestle with her while rolling away from 004. 004 moves to R5 and continues to
try to groom him as R5 continues to wrestle with 9T.
004 grabs R5 with both hands and wrestles him. R5
lies down and 004 again grooms him.

Here, and in other cases like this, play seems to function as a
mild reprimand. These contexts grade in intensity to the point
where they appear to be a form of 'punishment'.

Obviously, from the above examples, play occurs in many
situations that contextually appear to be kinds of parental behavior, involving the mother or adult animals in the group
other than the mother. The play that occurs may be initiated
by the older animals or the younger animals, but is usually
initiated by the older animals. Its function seems to vary
from that of a simple distraction tactic to a form of punishment that might be referred to as a 'veiled threat'. Generally, all function to change the behavior of another animal,
or at least appear to attempt to do so.

B.   *'Sibling Rivalry' and Other 'Jealousies'*

Some bouts of play involving adult animals appear to be
motivated by 'jealousy'. This is especially true of contexts
observed involving siblings.

7 July 1970 - *1627*: 7P grooms his mother 065. His
adult sister, OS, moves to him and bats his hands away

from her mother.   065 ground bats toward S, who cowers
from her mother.   7P resumes his grooming of 065.   OS
grabs and begins to wrestle 7P, who ignores her and
continues to groom his mother.   OS then wedges her body
between 7P and 065, thereby pushing 7P away, while she
begins to groom 065.

At times, siblings may appear to be jealous not only of
each other's interactions with their mother, but may also appear to be jealous of an older sibling's interaction with a
younger sibling.   Play appears in these contexts also.

7 May 1970 - *1005*:   Three-year-old 1C sits grooming his
two-year-old sister, 9V.   Their four-year-old sister,
M7, with her young infant clinging ventrally, moves to
the grooming animals and shoves 1C from 9V as she begins to wrestle him.   Their mother, DL, runs up and
bites M7, who sits grimacing and shrieking at DL while
9V stands screaming facing DL.   AC, DL's mother, runs
to and mounts 9T as DL cowers from her mother's approach.   AC moves to DL and touches her muzzle to DL's
muzzle briefly while rolling her hand slightly and
mouthing DL's muzzle.   When AC sits down, DL and 1C
groom AC.

While many of the contexts in which adult play appears do not
become so obviously complex, still a number of things should
be noted about this interaction: 1) the three-year-old male
(1C) was obviously in no way harming his two-year-old sister
(9V);   2) 9V was not an infant or yearling and, as such, supposedly should not be so jealously guarded by other siblings
with infants of their own; and 3) the mother (DL) obviously
did not regard the four-year-old female's (M7) play with the
three-year-old male as 'innocent pleasure'.

Adult animals were also observed to apparently be jealous
of young animals other than siblings who interacted with their
mothers.   Play may be seen here also.

9 July 1970 - *1632*:   Z9 sits huddled against 004. 004's
four-year-old son, R5, stalks toward Z9 who cowers and
moves from 004.   R5 sits touching 004 briefly and then
moves to and wrestles Z9.

Play appearing in contexts like this may be exhibited even
by very young infants.   The objects of the apparent 'jealousy'
are not necessarily always exclusively their mothers or siblings.

Another context in which 'jealousy' appears is in nongene-
alogical male/male dyads.   Adult males have been reported to
take an interest in young males, especially yearlings (cf.,
Breuggeman, 1973).   Adult play apparently motivated by jealousy may appear in such contexts:

17 July 1970 - *0800*:   UB grooms yearling W69F.   DS
moves to the pair, touching them briefly, then leans

toward UB, pulls him away from W69F and wrestles him.
UB ignores DS and continues to groom W69F.

C.   *Sexual Contexts*

Play behavior also appears in sexual contexts. While these
are usually somewhat confusing, they are worthy of note, especially those which may indicate endogenous as well as exogenous
factors as influencing adult play. Most of these situations
involved three- or four-year-old females.

12 November,1969 - *1722*: JS sits near estrous female
G8 who sits touching her mother. G8 moves to, rears
up, grabs and begins to wrestle her old adult sister
DL, who cowers from her and moves off. G8 then moves
to old adult female 076 and wrestles with her until
076, subordinant to G8, flees shrieking. JS moves to
and grooms G8. Infant 376 moves past and G8 grabs him
and places his penis in her mouth. 376 pushes at G8's
head and struggles away. G8 grabs and wrestles 376.

Not all play appearing in sexual contexts involved young adult
females.

5 November, 1969 - *0757*: JS sits touching estrous
female 076. WK, JS's brother, sits nearby. WK moves
to JS and stands presenting his hind quarters to JS.
JS, giving a rasping grunt, grabs WK's hind quarters
and while slowly rolling his head, begins to mouth at
the dorsal side of WK's tail. WK reaches back through
his legs and grabs at JS's penis. 076 moves away and
JS moves after her. WK moves to and wrestles TD, another adult brother.

Younger adult males are also seen to play in sexual contexts.

7 September, 1969 - *1522*: Estrous female N2 moves to
and presents her hind quarters to P7, who grabs and
wrestles N2. N2 ignores P7's play and continues to
stand in sexual present posture. P7 mounts N2 in a
series mount while she alternately grooms him.

These examples and other sexual contexts in which play was
observed seem somewhat different than other kinds of contexts
in which adult play was seen to occur. Most importantly, they
cannot clearly be assumed to occur with an attempt to manipulate the behavior of another individual given the manifest behavior of the animals involved. The observed contexts of adult
male/male play at the end of the mating season, which appear
almost chaotic in comparison with the contexts of other adult
play, reify this conclusion.

12 December 1969 - *0811*: DS mounts 1C, 1C grooms DS.
WK moves to and wrestles 1C as DS cowers from WK's
approach. CN moves to and presents to WK as 1C cowers

from CN's approach. WK pats CN's scrotum then lies
down and CN grooms WK. 1C moves to and is mounted by
EE, then EE and 1C wrestle. EE grooms 1C briefly and
then once more they wrestle. DS moves to and grooms
1C's sister, M7, who is mutually grooming with their
mother DL. 1C moves to and wrestles DS then 1C grooms
DS briefly before moving away. DS moves to and wrestles
1C.

Such a number of bouts, occurring in rapid succession, is not
generally characteristic of adult play.

D. Agonistic Contexts

Adult play was also seen to occur in a number of contexts
with agonistic behavior. The sequence of these contexts was
not that of play grading into agonistic behavior, but rather
that of play following an agonistic encounter. One such context involved myself.

1 April 1970 - *1529*: EE stands lunging and grunt
threatening me. WK moves to me, places his hand on
my boot, then turns and presents his hind quarters
to me. WK then turns again, grasps my pants leg with
one hand and begins to gently and slowly wrestle with
me while mouthing my leg. EE cowers and moves off.

Such contexts most frequently involved young animals.

13 April 1970 - *1646*: 1C lunges and ground bats at
8T who jumps away, cowering then runs to and wrestles
with WK. 8T then mounts WK and then wrestles with him
again.

During both of these encounters, WK was the highest ranking
male in group F. These contexts seem essentially similar to
those in which adult caretakers 'protected' an infant or yearling. Not all involving adult play are, however.

1 June 1970 - *1744*: 004 lunges at her son R5 who
cowers from her and then open mouth faces her younger
sister 9T, who runs shrieking to 004 and is groomed
by 004. R5 grabs his infant brother 309 and wrestles
with him roughly. 004 charges R5 who again cowers
but turns and open mouth faces 9T who flees to 004's
side. R5 once more grabs and wrestles 309 roughly
until 309 can move away from him.

In those contexts where play appeared to function as a form
of redirected activity, infants or other very young animals
were frequently the objects of this redirection, and most frequently they were the younger siblings of offspring of the animal who had been dominant in the agonistic encounter. This was
not always so, however, especially in cases such as the following:

12 June 1969 - *1522*: R5 tries to groom his older sister, WT. WT moves away with her infant on her ventrum. R5 moves to her and again tries to groom her and once more WT moves away with her infant. R5 runs to WT, grabs and wrestles with her, while giving rapid, mock neck bites repeatedly.

While there were no overt elements of agonistic behavior in this context, the play itself seems to function as a kind of agonistic behavior. This is clearer in the following context, involving two adult males:

21 July 1969 - *1006*: DS sits by himself on a rock. JS moves to DS, who grimaces at JS, but remains sitting on the rock. JS places one arm about DS's neck and begins to wrestle him while giving repeated mock bites at DS's side until DS cowers and moves from the rock. JS sits on the rock and grooms himself.

Adult play was observed to occur in other contexts (such as when a male encountered a male from another group), although more rarely than in one of the previously mentioned contexts. Due to the infrequency of adult play, any one of the above kinds of contextual examples may be subject to disclaimer. Taken as a whole, however, they imply that adult play behavior is in no way a meaningless or afunctional behavior.

VII. THE QUESTION OF FUNCTION RECONSIDERED

It appears obvious that even among adult rhesus, play behavior is a multifunctional behavior. The contexts in which it occurs suggest it frequently serves a number of immediate social functions which may be categorized as forms of low arousal, relatively nonthreatening social manipulation. Because young animals were sometimes seen to initiate this form of play with adults, it is most reasonable to assume this to be a potential generalized function of play behavior throughout the life cycle of rhesus monkeys. In that the play of the young is frequently much more varied in kind, form and manner than that of adults, it is then unlikely that this is the only function of play in rhesus.

VIII. CONCLUSIONS

(1) Play behavior within a species can be multifunctional.

(2) Closer attention must be paid to the life cycle context as well as the immediate behavioral context in order to understand the potential functions of play within a species.

(3) Play in adult rhesus may be identified by the same basic characteristics as play in the young. It is, however, usually more simple in form and kind.

(4) Generally, the older a rhesus adult, the less frequently play occurs, although all adult animals in the group were seen to play except for old adult, low ranking females. Most play of adults is with animals of two years of age or younger. The older animals were seen to initiate this play more frequently than the younger animals.

(5) There are few apparent sex differences in the kind of play exhibited by adult rhesus and there were no sex differences in the frequencies of play exhibited in the category of "old adult".

(6) Higher ranking adults were seen to play more frequently than lower ranking adults. Usually the animals they played with were younger animals of their own genealogy.

(7) Most mothers were seen to play at some time with, usually their youngest, offspring.

(8) The contexts in which adult play in rhesus was observed may be categorized under four general headings: a) mother/offspring and other parental dyads; b) 'sibling rivalry' and other 'jealousies'; c) agonistic contexts; and d) sexual contexts, the latter being a context which suggests that endogenous (i.e., hormonal influences) as well as exogenous (i.e., social and biological relationships and the concomitants of those relationships) factors may affect the occurrence of play behavior.

(9) Most bouts of play involving adults, while serving a variety of immediate functions, functioned overall as a low arousal, relatively nonthreatening form of social manipulation.

(10) While it is highly unlikely that this is the only function of play in rhesus, social manipulation is likely a potential function of play throughout the life cycle.

REFERENCES

Aldis, O. "Play Fighting". Academic Press, New York (1975).

Bramblett, C. A. "Patterns of Primate Behavior". Mayfield Publishing Co., Palo Alto, California (1976).

Breuggeman, J. A. Parental care in a group of free-ranging rhesus monkeys (Macaca mulatta). Folia primat. 20, 178-210 (1973).

_____ "Ontogeny of Social Relations in a Group of Free-Ranging Rhesus Monkeys (Macaca mulatta Zimmerman)". Ph.D. Dissertation, Northwestern University, Evanston, Illinois (1976).

Huizinga, J. "Homo ludens: A Study of the Play-Element in

Culture". Beacon Press, Boston (1950).

Loy, J. Estrous behavior of free-ranging rhesus monkeys *(Macaca mulatta)*. *Primates 12*, 1-30 (1969).

Sade, D. S. An ethogram for rhesus monkeys. I. Antithetical contrasts in posture and movement. *Am. J. phys. Anthrop. 38*, 537-542 (1973).

Smith, W. J. Message, meaning, and context in ethology. *Am. Nat. 99*, 405-409 (1965).

THE QUESTION OF FUNCTION:
DOMINANCE AND PLAY

Donald Symons

Department of Anthropology
University of California, Santa Barbara
Santa Barbara, California

*"Each woman presents a different problem, or, alterna-
tively, each woman presents the same problem" (Stephen Potter,
"Woo Basic").*

I. INTRODUCTION

Each behavior pattern has the same function, to contribute
to reproductive success (Hinde, 1975), but the interesting
problem is what specific contributions to this goal a given
behavior pattern has been designed to make. Function is the
most vexing and intractable issue in the study of primate
behavior. Aside from the inherent difficulty of the problem,
this circumstance results, I believe, from a general failure
to consider carefully what "function" means in biology, from
the application of social science concepts of function to
nonhuman primates, and from the attempt to find a moral order
in nature. The question of function in primate behavior --
especially the function of play and aggression -- thus raises
issues not generally raised by other kinds of behavioral
questions. In this essay, these issues and their interrela-
tionships are examined, in part, by analyzing a specific
hypothesis: that play functions to establish a dominance
order.

II. FUNCTION IN BIOLOGY

Species-typical behavior in a natural habitat has both
proximate and ultimate causes (Mayr, 1961). The failure to
distinguish clearly between them results in fruitless controversy (see Lehrman, 1970). The proximate, or immediate,
causes of a behavior pattern are the interactions of: the
particular complement of genes the organism possesses; the
particular environmental conditions it encountered during ontogeny; the presence of particular endogenous and exogenous
stimuli. The ultimate, or evolutionary, causes of a behavior
pattern are specific selective forces in ancestral populations. Organisms are entities designed to promote, not their
own survival, but the survival of their genes. In this context, "designed to promote" is equivalent to "whose purpose
is to promote", or "whose goal is to promote", or "whose function is to promote". Functional design implies the existence
neither of a designing entity nor of genetic, supernatural, or
organismic consciousness. The ultimate cause of functional
design is the cumulative action of mindless evolutionary processes: chance variation and differential reproduction.[1]
Since selection eventually eliminates nonfunctional behavior (Hinde, 1975), a species-typical behavior pattern in a
natural habitat can be expected to have at least one beneficial effect that may properly be called its function, goal, or
purpose (Williams, 1966), although the same behavior pattern
can have other effects -- beneficial, harmful, and neutral --
on an animal's reproductive success.[2] Williams (1966) and
Hinde (1975) argue that the study of function has been impeded
by the assumption, often implicit, that all beneficial effects
are functions. "Function" ultimately refers not to the effects of an animal's behavior but to the basis of differential
reproduction among its ancestors. The statement that an

[1]In biology and in the social sciences, "function" sometimes is used in analogy with mathematical usage to indicate
that something depends on and varies with something else.
Klein (personal communication) proposes the phrase "proximate
function" for this meaning and "ultimate function" for the
evolutionary meaning.

[2]Hinde (1975) suggests three possible explanations for the
existence of a nonfunctional character: (1) it may be a byproduct of a functional character; (2) it may typically be
functional, but occur occasionally in a nonfunctional context;
(3) it may represent an adaptation to environmental circumstances that no longer exist, and, hence, be in the process
of being lost.

effect (E) is a function of a behavior pattern (B) implies
that B was fashioned by natural selection to produce E. B may
fortuitously produce other beneficial effects, but these
chance effects reveal nothing about evolutionary processes.
    That an animal's behavior benefits the social group, popu-
lation, species or ecosystem of which it is a part is not evidence that such behavior functions to produce these benefits;
group benefits can arise as incidental effects - statistical
summations of individual adaptations (Richerson, 1977; Williams, 1966). Williams (1966:4) argues, on the basis of par-
simony, that "...adaptation is a special and onerous concept
that should be used only where it is really necessary. When it
must be recognized, it should be attributed to no higher a
level of organization than is demanded by the evidence." Thus
the simplest form of natural selection, that of alternate
alleles in Mendelian populations, should be assumed to be sufficient to explain the function of behavior unless the evidence indicates that it is inadequate. Many scholars have
argued that functional hypotheses that assume the existence of
"group selection" are not only unparsimonious but are markedly
inferior to individual and kin selection hypotheses in accounting for the available data on animal behavior (e.g.,
Dawkins, 1976; Ghiselin, 1974; Klein, 1966; Lack, 1966; Tinbergen, 1967; Williams, 1966). Furthermore, the conditions
under which group selection could, in theory, become more po-
tent than individual selection appear to be unlikely to occur
in nature (Maynard Smith, 1976).
    While biologists differ as to the criteria for determining
the function of behavior, perhaps all would agree that the
first step is to observe behavior, preferably in a natural environment, to describe it accurately, to measure it where possible, and to determine the contexts in which it occurs
(Hinde, 1975; Tinbergen, 1967). Four kinds of evidence have
been considered relevant to determining function: the correlation of character variation with variation in reproductive
success, experiment, convergence and divergence, and design.

A. *The Correlation of Character Variation with Variation in*
    Reproductive Success

    Hinde (1975) argues that correlating individual variation
in the expression of a character with variation in reproductive success can demonstrate function in a "strong sense".
There are, however, a number of problems with this approach.
First, as Hinde himself points out, observed variation in a
character may not have a genetic basis. Second, as Hinde also
notes, variation in reproductive success may actually be
caused by variation in a second character with which the

character being observed is correlated. Third, in a natural environment, variations of major evolutionary significance may be too small to be detected (Tinbergen, 1965); the stronger the selection pressures maintaining a character, the less variable it is likely to be. Fourth, the selection pressures maintaining a character may differ from the pressures that molded it (Tinbergen, 1965, 1967). This point is exemplified below.

B. Experiment

While experimentation is not always efficient, possible, or necessary, it can sometimes provide otherwise unobtainable evidence about a character's effects (Hinde, 1975; Tinbergen, 1967). Ethologists, through brilliant field experiments, have circumvented the problems presented by minimal naturally occurring variation in the expression of a character and unsuspected covariations among characters. For example, the mottled pattern of Black-headed gulls' eggs strongly suggests selection for camouflage. This hypothesis was tested by comparing the rate of predation -- away from the nest -- on normal and experimental, solid color eggs: the experimental eggs were, indeed, far more frequently preyed upon (Tinbergen, 1967). Even when natural selection is presently operating to maintain a character through an experimentally demonstrated effect, however, stabilizing selection pressures may differ from the pressures that molded the character. For example, the experiment just cited was conducted initially in Black-headed gulls' nests, but it turned out that parent gulls sit better on normal than on solid color eggs, and the observed difference in rates of predation could have been caused by the difference in effective sitting and not by the difference in camouflage (Tinbergen, 1967). It is possible, then , that selection maintains normal egg coloration both because such eggs are preyed upon less frequently and because they are sat upon more effectively; but this does not mean that selection for effectiveness as a sitting stimulus played a part in the evolution of the normal egg coloration.

C. Convergence and Divergence

Although every lineage has a unique history, function can nevertheless be illuminated by the ways of life common to distantly related species that have independently evolved similar characters (convergence) and by the contrasting ways of life of closely related species that have evolved different characters (divergence). For example, Cartmill (1974)

rejects the traditional view that typical primate characters
(nailed digits, opposable thumbs and first toes, close-set
eyes) were originally arboreal adaptations because nonprimate
arboreal mammals do not both diverge from allied terrestrial
species and converge toward primate-like specializations. The
independent evolution of primate-like characters among certain
marsupials, chameleons, cats, and owls indicates, according
to Cartmill, that primate specializations were designed originally for visually directed predation.

D. Design

Williams (1966:209) argues that "The demonstration of a
benefit is neither necessary nor sufficient in the demonstration of function, although it may sometimes provide insight
not otherwise obtainable. It is both necessary and sufficient
to show that the process is designed to serve the function."
If an effect can be adequately explained as the result of
physical laws or as the fortuitous byproduct of an adaptation,
it should not be called a function; function implies design:
"The decision as to the purpose of a mechanism must be based
on an examination of the machinery and an argument as to the
appropriateness of the means to the end. It cannot be based
on value judgments of actual or probable consequences" (Williams, 1966:12).
Design is revealed in the precision, economy, and efficiency with which goals are attained, but adaptations do not
necessarily approximate the theoretically achievable uncon-
strained ideal: design is constrained by the specific history
of a lineage, by the randomness of mutation with respect to
fitness, by opposing selective forces, and by the necessity to
compromise the selective demands of other adaptations (Williams, 1966). Williams points out that analogies between biological mechanisms and artifacts designed to serve human purposes -- the lens of an eye and the lens of a camera -- sometimes are useful in elucidating design. I have suggested that
certain aspects of the learning and practice of sports skills
may provide insights into the design of animal playfighting
(Symons, 1978).
According to Williams, the scientific study of design --
the demonstration of appropriateness of means to ends -- is in
a primitive state of development, in part because design so
often is obvious; a sophisticated science is not needed to
elucidate the visual purpose of the vertebrate eye. Hull
(1974:123) suggests that human intuition is a powerful tool in
recognizing the goals of biological mechanisms because "we
ourselves are teleological systems and arose through the same
selection process as these other teleological systems, in part

because of our ability to recognize and to respond appropriately to them as coherent entities."[3] Function appears to be so easily recognized that when, as is frequently the case in the study of behavior, intuition fails or different people reach different intuitive conclusions, progress is difficult; the absence of necessity has resulted in the absence of invention of the requisite scientific tools. Consequently, "Just because hard evidence is so difficult to obtain, it has become respectable to speculate about the function of behaviour in a manner that would never be permissible in studies of [immediate] causation" (Hinde, 1975:13).

That effect, but not function, can be demonstrated experimentally is not grounds for equating function and beneficial effect. The demonstration of a benefit may in some cases provide insight into function; in other cases -- for example, when normal development is shown to depend on the occurrence of a given behavior pattern -- the effect is necessarily of interest to the student of proximate causation but is not necessarily of interest to the student of ultimate causation; and many beneficial effects are irrelevant to any significant biological problem. That function is often equated with beneficial effect may account for a widespread conviction that most behavior patterns serve many functions. Multiple functions do exist (Beer, 1975), but circumstances in which selection can simultaneously maximize two or more functions probably occur infrequently, and multiple functions can generally be expected to entail compromise in design. Sexual dimorphism in the design of pelvic bones, not the fact that a human infant passes through its mother's pelvis during birth, demonstrates that a function of the pelvis of the human female is to accommodate an infant's head.

[3]*If human purposiveness is, itself, an adaptation, evolutionary purpose and human purpose are more than analogous. The conscious purposes of humans in a natural environment presumably are the outcomes of selection to maximize genetic representation in succeeding generations, just as the biological processes not represented in consciousness are. Unlike other biological processes, however, consciousness not only gives the appearance of anticipating future circumstances, but actually does so: imagined effects can cause behavior. Thus, in a natural habitat, human purposes are not only like evolutionary purposes, in large part they are evolutionary purposes. A mindless process has created mind in its own image.*

E. *The Influence of Social Science Concepts of Function in Primatology*

Maynard Smith (1976) suggests that the vehemence with which proponents of individual and kin selection have pursued the case against group selection results not only from their belief that group selection is an insignificant evolutionary force, but also from "their conviction that group selection assumptions, often tacit or unconscious, have been responsible for the failure to tackle important problems" (p. 277). Primatologists such as Clutton-Brock and Harvey (1976) and Popp and DeVore (in press) maintain that implicit group selection assumptions have retarded the understanding of nonhuman primate behavior, and they propose alternate explanations of behavior based on inclusive fitness theory. Since the early 1960s, the influence of group selection theories in biology has declined as these were made explicit and testable; but many primatologists who disavow group selection continue to propose functional explanations of behavior that appear to imply group selection. Perhaps this is because primatology does not derive its concepts of function from biology, but rather from the social sciences. Primatology may be especially likely to incorporate social science concepts into interpretations and explanations of behavior because: (1) primatologists frequently are trained in, and find employment in, departments of social science; (2) many primatologists believe that the study of nonhuman primate behavior can provide insight into human behavior and human evolution; (3) as Thompson (1976) points out, until recently many primatologists owed their considerable success in obtaining research funds to their ability to convince granting agencies that studies of nonhuman primate behavior have important implications for understanding and solving human problems.

Contrary to Sahlins (1976), who argues that it will be necessary to take a "superorganic" perspective in order to understand animal social organization, I believe that the superorganic perspective -- the concept of a superindividual system with needs of its own -- perhaps more than group selection assumptions, has been responsible for errors in the study of function in primate behavior. The popularity in primatology of such concepts as "socialization" and "role", for example, which according to Rowell (1975) and Hinde (in press), respectively, are misleading when applied to nonhuman primates, reflects the influence of a superorganic, functionalist social science. This point can be clarified by a brief consideration of social science concepts of function. Only the views of Radcliffe-Brown and Malinowski are referred to explicitly, but the discussion is intended to be generally applicable.

Radcliffe-Brown believed that culture and society function to promote stability and cohesion within the existing system of human social relations; that is, culture and society promote the material well being of the collective (Hatch, 1973). Radcliffe-Brown's functionalism is explicitly based on the analogy -- which appears to have its origin in Plato's *Republic* -- between a human society and an organism:[4] "The concept of function applied to human societies is based on an analogy between social life and organic life" (Radcliffe-Brown, 1935: 394). A society has various "necessary conditions of existence"; the contribution each component in the society makes to creating these conditions is the component's function, and the social system thus has a "functional unity". In the social sciences, most functionalists have shared Radcliffe-Brown's assumption that societies are harmonious wholes (Jarvie, 1964). Martindale (1965a:ix) writes: "Theoretically, [the functionalist point of view] consists in the analysis of social and cultural life from the standpoint of the primacy of wholes or systems." Spencer (1965:14) writes: "By their very nature, human societies, suggesting as they do the holistic organism, are to be seen in their context of balanced, integrated entities." According to Martindale (1965b: 154), two basic propositions guide functional analysis: "(1) the social system itself is the prior causal reality and (2) any given part or component of the social system is best conceived as in functional interrelation with other components of the whole."

Malinowski used "function" in a number of different and imprecise ways that he himself did not explicitly distinguish (Hatch, 1973). In most of his writing, Malinowski argued that culture and society serve individual, not collective, ends: "he conceived culture as an instrument serving man's biological and psychological needs: the end served by function is individual imperatives, and not social stability and cohesion as Radcliffe-Brown held" (Hatch, 1973:320). But Malinowski's views were not so individualistic as they first appear. For one thing, he sometimes did use "function" to refer to the

[4]*This analogy is satirized by Jonathan Swift in* Gulliver's Travels: *"whereas all writers and reasoners have agreed that there is a strict universal resemblance between the natural and the political body", wrote Lemuel Gulliver, "can there be anything more evident than that the health of both must be preserved, and the diseases cured by the same prescription?" Nothing could have been more evident to certain professors in the Grand Academy of Lagado, who proposed that physicians should daily feel the pulses of every senator and then prescribe appropriate medical treatment in order to cure the diseases of head and heart endemic to legislative bodies.*

contribution a part of culture makes to the persistence and
integration of the community as a whole (Hatch, 1973: Malinowski, 1939). For another, Malinowski's argument that culture
serves the biological and psychological needs of individuals
conceals an indirect group benefit. In *Sex and Repression in
Savage Society* (1927), Malinowski argues that human culture
and animal instincts are functionally identical: "Thus it can
be said without exaggeration that culture in its traditional
bidding duplicates the instinctive drive" (Malinowski, 1927:
209-210). But to Malinowski, instinctive drives have nothing
to do with Darwinian fitness: an animal's instincts function
to promote the survival of its species, and similarly, although culture functions to satisfy the biological and psychological needs of individual human beings, these needs themselves have a function, to promote the survival of the human
species. That culture ultimately serves a group benefit is
made clear in Malinowski's explanation of the human family.
The family serves two functions: "it has to carry on propagation of the species and it has to insure the continuity of
culture" (Malinowski, 1927:261).

When applied to nonhuman animals, social science concepts
of function are misleading in several respects. First, Radcliffe-Brown, Malinowski and more recent functionalists share
the assumption that in some sense society and culture have an
autonomous, superorganic existence. The appearance of auton-
omy is made possible primarily by language; since nonhuman
animals lack language, they essentially lack society and culture in the human sense.

Second, while the organismic analogy may be useful in the
social sciences, from an evolutionary perspective, a nonhuman
animal cannot be considered to be analogous to its group: an
organism's cells, tissues, and organs are 100 percent cooperative because they are 100 percent related genetically. A pan-
creas cell contributes to the continued existence of its genes
not by striking out on its own or by forming an alliance with
a spleen cell in order to take over the body, but by staying
home and cooperating with the other body cells in their effort to assist a few of the gametes in building new organisms
in which their genes can survive. But the individual members
of an animal group are imperfectly related genetically,
achieve reproductive success at one another's expense, and
compete for the same resources. Modern functionalist doctrines generally compare human society with animal physiology
-- which is assumed to have homeostatic, "system-maintaining"
functions -- not with animal behavior. Since the survival of
an animal's genes necessitates the temporary survival of the
animal, it is easy to see how social scientists arrived at the
erroneous conclusion that an animal's physiology is adapted to
promote its survival, but it strikes me as unlikely that the

study of animal behavior *per se*, especially when informed by
evolutionary theory, would ever have suggested the idea of a
"functional unity" or given rise to such concepts as "role"
and "socialization". It seems likely that, owing to certain
peculiarities of Western intellectual and religious traditions, human society was conceived first in analogy with an
organism and then animal society was conceived in analogy with
human society.

Finally, although social science concepts of function appear to imply group selection, in an important sense they are
not selectionistic at all. As discussed above, in biology,
function is essentially historical: ultimately it refers, not
to effects themselves, but to a creative process -- differential reproduction -- that produced the effects. In social
anthropology, on the other hand, functionalism is essentially
ahistorical: it arose largely because Malinowski and Radcliffe-Brown rejected the practice of explaining human society
by historical conjecture and maintained that society must be
understood with reference to the present (Hatch, 1973; Jarvie,
1965). Although a definite view of historical processes is,
in fact, implicit in functionalism (Hatch, personal communication), Malinowski, Radcliffe-Brown, and more recent functionalists avoid historical analysis and emphasize the operation
of existing social systems.[5] Most frequently, the functions
of a component are defined as its objective, observable,
system-maintaining effects (Hempel, 1959; Holt, 1965; Merton,
1957). Hence, social science concepts of function differ fundamentally from biological concepts.

Functionalism has been widely criticized, and many non-
functionalist positions exist in the social sciences (see,
e.g., Barth, 1966; Boissevain, 1974; Jarvie, 1973; Leach,
1961; Martindale, 1965b; Richerson, 1977); indeed, functionalism has had relatively little influence in economics (Krupp,
1965). These criticisms and alternate positions need not be

[5] *There is no hint of consensus in the social sciences
about basic units of society, corresponding to genes, about
basic creative processes by which social forms are generated,
corresponding to natural selection, about the sources of innovation, corresponding to mutation, or, indeed, about the
relationship between human social life and individual reproductive success. Society and culture are variously considered: to promote their own survival as coherent systems;
to consist of elements, each of which promotes its own survival; to promote the survival or satisfaction of individual
human beings, groups of human beings, ecosystems, or genes
(see, e.g.,* American Psychologist 31: *341-384, 1976; Campbell, 1975; Cloak, 1975; Dawkins, 1976; Ruyle, 1973; Sahlins,
1976; Wilson, 1975).*

reviewed here, since my purpose is not to analyze the contribution of functionalism to the study of human beings; but a functionalist perspective is highly misleading in primatology. Functionalist concepts: (1) imply the existence of a super-individual system with needs of its own, which, in turn, implies that behavior is substantially influenced by language-based traditions inherited in Lamarckian fashion; (2) imply that groups constitute harmonious wholes; and (3) are essentially nonhistorical, referring to effects, not to processes. The influence of social science concepts of function may, in part, explain present tendencies in primatology to equate function and beneficial effect, to give functional explanations that seem not to be genuinely historical, and to treat social integration as if it, and not reproductive success, were the ultimate goal of behavior.

F. The Function of Play

If function in general is difficult to demonstrate, the function of play is still more problematical, since a defining characteristic of play is frequently considered to be lack of immediate benefit (e.g., Fagen, 1974); thus, by definition, the function of play becomes obscure. Such definitions have been criticized because play may, in fact, provide immediate benefits (e.g., Beach, 1945; Ghiselin, 1974) and because the benefits of any behavior pattern, even eating, lie in the more or less distant future (Millar, 1968); but however play is defined, its functions are not immediately obvious.

As nonhuman primate play rarely has been described in anything like the detail that is customary in descriptions of slower-paced, more stereotyped activities, evidence about function has not been widely available. Consideration of the list compiled by Baldwin and Baldwin (1977) of functions that have been variously proposed for primate play indicates that most functional hypotheses are informed guesses about benefits that primates might derive from playing. Most frequently, these benefits are considered to be the acquisition of social rather than physical skills.

That monkeys develop abnormally when deprived of early peer contact is sometimes cited as support for functional hypotheses of play (e.g., Poirier and Smith, 1974). As has often been pointed out (e.g., Bekoff, 1976; Dolhinow and Bishop, 1970), however, such deprivation experiments cannot demonstrate that play affects development, since monkeys are not selectively deprived of play, but of all peer contact. "There does not appear to be one case in which it has been demonstrated that play experience is *essential* for socialization to occur" (Bekoff, 1976:499). Indeed, the existence of

apparently normal, nonplaying squirrel monkeys (Baldwin and
Baldwin, 1973, 1974, 1977) indicates, at least in this species, that play is not essential.

Recently, Baldwin and Baldwin (1976) and Oakley and Reynolds (1976) have altered play frequencies in captive primate
groups by manipulating feeding patterns. Such techniques may
demonstrate eventually that play affects development, but, as
discussed above, function is not thereby demonstrated. For
example, play may have evolved to serve some nondevelopmental
function and may have gradually become essential to normal
development. Wilson (1975) suggests that learning processes
may become incorporated into the development of a species-
typical behavior pattern as an incidental effect, or byproduct, of selection for neural economy. If certain stimuli are
reliably present in the environment of developing organisms,
those immature animals that, as a result of a chance mutation,
require the presence of these stimuli to develop normal behavior will be more fit than those that do not require the presence of these stimuli if the neural mechanisms underlying the
former type of development are more simple, and, hence, more
energetically economical, than the mechanisms underlying the
latter type of development. This is analogous to the evolutionary loss of the ability to synthesize a biologically
necessary chemical when the chemical becomes reliably and uniformly present in a species' diet: selection favors individuals who economize on energy expenditures by eliminating
superfluous biochemical processes. As the ability to synthesize the chemical is lost, its presence in the diet becomes
necessary for survival, and the chemical becomes a "vitamin"
(Wilson, 1975). Thus, "Even were play a necessary condition
for normal development, as may well be the case, it does not
follow that animals play for that reason. A brainless sala-
mander embryo will not develop eyes, but that does not mean
that salamanders have brains in order to provide them with
those parts" (Ghiselin, 1974:260).

Fagen's (1976) paper, "Exercise, play, and physical training in animals", is an example of a functional hypothesis
about play that is supported with relevant data. Because play
entails vigorous exercise in apparently nonfunctional contexts, Fagen proposes that it functions as physical training.
On the basis of a review of the literature on the known effects of training, Fagen argues in essence that many features
of play appear to be designed to achieve this goal, and he
derives specific predictions based on this hypothesis about
the distribution, ontogeny, and detailed structure of play.
Symons (1974) suggested that rhesus monkeys are unlikely to
have the sorts of experiences during aggressive play needed to
learn to interpret or use agonistic signals, and, therefore,
concluded that, contrary to suggestions in the literature,

play is not designed to promote the learning, refining, or
practicing of communication skills.   Theory and data already
available in the literature can, thus, shed light on functional hypotheses.   To illustrate the usefulness of a rigorous
approach to function, in the following section the available
evidence is used to analyze one of the most prevalent hypotheses in the literature on nonhuman primate play.

III. DOMINANCE AND PLAY

    Stated with varying degrees of confidence and explicit-
ness, that play functions to establish a dominance order is
perhaps the most common functional hypothesis in the literature on nonhuman primate play (Carpenter, 1934; Dolhinow,
1971; Fagen, 1976; Gottier, 1972; Hall, 1965; Harlow, 1969;
Harlow and Harlow, 1965; Jolly, 1972; Poirier, 1970, 1972;
Poirier and Smith, 1974; Rosenblum, 1961; Suomi and Harlow,
1971).   To my knowledge, this view has been questioned in
print twice.   Meier and Devanney (1974:293) imply disagreement
with the dominance hypothesis: "...in contrast to the common
supposition that play is an avenue for the immature animal
'to learn social behaviors', the ontogeny of play occurs in a
social context within which the infant is already responding
as an established member."   Aldis (1975:173) directly chal-
lenges the dominance hypothesis, arguing in essence that although the establishment of rank order relationships may be an
effect of play, play does not appear to be designed to achieve
this goal: "...the influence of play on rank order relationships is probably only adventitious: it is difficult to believe that the survival value of play depends primarily on its
effect on rank order relationships."   A brief review of recent
factual and theoretical contributions to the study of dominance and play will provide a framework for evaluating the
dominance hypothesis.

A. *Aggression and Dominance*

    Recent developments in evolutionary theory and field stud-
ies of animal behavior warrant three conclusions about intraspecific aggression: (1) The ultimate cause of intraspecific
aggression is competition for scarce resources (Lack, 1969;
Marler, 1976).   (2) Selection among alternate alleles in
Mendelian populations can account for naturally occurring patterns of aggression and for the evolution of agonistic signals
(Dawkins, 1976; Maynard Smith, 1972, 1974; Maynard Smith and
Price, 1973; Parker, 1974).   (3) Aggression is a far more

important cause of mortality among free ranging animals than
has generally been believed (Marler, 1976): mortality results
both from fights and from the withholding of resources by territoriality and dominance which Marler (1976) calls "quiet
violence".

Serious fights occur as infrequently as they do because
fighting entails risk and because, even when successful,
fighting may be an ineffective competitive strategy. For example, two anubis baboon males fighting over an estrous female
risk not only injury but insult: while they battle, a third
male may abscond with the prize (Popp and DeVore, in press).
Natural selection favors willingness or desire to fight only
when benefits typically exceed costs in the currency of reproductive success; but it takes at least two to make a fight
(as distinct from an unreciprocated attack). When individuals
can assess relative fighting abilities with a high degree of
accuracy, selection generally favors discretion on the part of
the probable loser. Detailed discussions of nonhuman primate
aggression and dominance from this perspective are becoming
available (Clutton-Brock and Harvey, 1976; Popp and DeVore, in
press; Symons, 1978) and only conclusions will be presented
here.

Most nonhuman primates are long lived, group living animals with excellent memories and acute sensitivities to the
opportunities and constraints sociality entails. The abilities to recognize group members individually, to remember the
results of past conflicts, and to assess relative competitive
abilities allow individuals to avoid fights that they are unlikely to win, or that entail too great a risk, and, hence,
rampant violence is rare; usually, A is observed to aggress
against B, but B flees or submits rather than retaliating,
is alert to A's moods and whereabouts, and avoids provoking A.
Dominance/subordinance relationships and hierarchies are the
outcomes of strategic individual compromises, the consequences, not the causes, of the reluctance of individuals to aggress (Clutton-Brock and Harvey, 1976; Dawkins, 1976; Hausfater, 1975; Popp and DeVore, in press; Symons, 1978;
Williams, 1966). "Hierarchies occur because competitive ability inevitably varies between individuals and because less
successful animals learn not to contest access to encounters
where they are unlikely to win thus saving time and energy"
(Clutton-Brock and Harvey, 1976:216-17). The conception of a
dominance hierarchy as a reified entity in which animals find
their place, rather than as an outcome of compromises among
competing, self-interested individuals -- the view that primate "society" is the cause rather than the effect of primate
behavior -- may be partly responsible for the widespread intuitive appeal of the hypothesis that play functions to establish rank order.

While a substantial body of evidence demonstrates that,
on the average, high ranking animals have greater access to
scarce resources than low ranking animals do (Wilson, 1975),
an evolutionary view of behavior makes this conclusion inevitable, whether or not differentials are large enough to be
reliably measured, since only in this circumstance could selection have favored the risk taking and energy expenditure
required to achieve and to maintain rank. In the absence of
resource competition, intraspecific aggression and, hence,
dominance, would not exist. Clutton-Brock and Harvey (1976)
point out that although there are far more unforced than
forced retreats by the subordinate member of a pair, it is
nonetheless the infrequent attacks or threats by the dominant
member that cause unforced retreats; hence, the dominant
animal really controls the relationship. Delgado (1974:258)
notes that reducing the frequency of aggression by the dominant male of a rhesus monkey group through intermittent
radio stimulation of the caudate nucleus "may abolish its
dominance and change the hierarchical structure of the whole
group."

physio logical

That aggression and the fear of aggression that is the
basis of dominance relations result ultimately from resource
competition does not mean that resource competition is invariably, or even usually, the proximate cause of aggression.
For example, the ultimate cause of most intermale violence
among free ranging rhesus monkeys appears to be breeding
competition, yet field workers generally report little overt
or obvious fighting over sexual access to an estrous female
(Symons, 1978). "Dominance" seems to be a natural category,
recognized by animals as well as by observers, and accession
to social position rather than access to material resources
frequently may be the immediate goal: "In the long run, position guarantees reward, but in the short run, position itself
is the reward" (Washburn and Hamburg, 1968:473).

B.   *Fights and Playfights*

Fighting is the most intense form of competition. Although most agonistic interactions consist of signals of
threat and submission, symbolic aggression is effective only
because the possibility of escalation exists: among species
capable of inflicting injury, wholly symbolic agonistic con-
tests cannot, in theory, evolve, since in a population of
symbolic aggressors, selection would favor a mutant that

harmed its opponents (Maynard Smith and Price, 1973).[6] During a fight, each combatant attempts simultaneously to inflict injury and to avoid being injured. Since fighting entails serious risk, it probably occurs only when combatants are so closely matched that the outcome cannot be predicted. Once a fight has begun, it is to the loser's advantage to detect defeat as quickly as possible in order to flee or submit with minimal injury. As the loss of a fight is unambiguous evidence of inferior competitive abilities, the loser is not likely to contest access to resources or social position with the winner soon. Fights can, thus, be important determinants of rank order relationships.

Nonhuman primate playfighting appears to be structurally similar to fighting: dominance and subordinance are absent, and participants simultaneously attempt to bite or to hit without being bitten or hit (Aldis, 1975; Owens, 1975a,b; Symons, 1978). The structural similarities of fighting and playfighting probably inspired the hypothesis that outcomes of playfights determine rank order. In evaluating this hypothesis, then, it is important to consider as well some of the ways in which fighting and playfighting differ.

The most fundamental structural difference between fighting and playfighting is that the movements -- bites or hits -- that can inflict injury during a fight occur at noninjurious intensities during playfights. While it is not clear to what extent immature animals are capable of actually inflicting injury, anyone who has seen a playfight turn into a fight can testify that juveniles are quite capable of inflicting pain and that playfighting occurs at less than maximum intensity. Playfights, unlike fights, are common between unevenly matched animals, are often initiated by an animal that is obviously outclassed, and do not terminate as a rule when it becomes clear which is the superior playfighter. Among rhesus monkeys (Lindburg, personal communication) and vervets (Lancaster, personal communication), juvenile males may playfight vigorously with fully adult males more than twice their size. The film "Rhesus Play", which John Melville Bishop and I recently completed, contains a sequence in which a large juvenile male initiates a series of fast-paced playfights with an adult male and is thoroughly bested by the adult in every one. Because fighting entails serious risk, ritualized signals have evolved by which animals communicate aggressive intent and, thereby,

---

[6]*Fagen (personal communication), however, recently found that the hawk/dove model is structurally unable to encounter nonrandomness: if in the population like encounters like with greater than chance frequency, there exists a cost/benefit ratio for which the symbolic aggressor -- dove strategy -- is evolutionarily stable.*

avoid fighting; because playfighting does not entail serious
risk, ritualized signals have evolved by which animals communicate intent to play and by which they solicit play and
thereby facilitate playfighting.

Competition over resources, which promotes fighting, in-
hibits playfighting. Meier and Devanney (1974) and Symons
(1978) note that an outbreak of aggression in a rhesus
monkey group can put a stop to play for hours, even when the
players are far from the aggressive interaction. In an established laboratory group of infant rhesus monkeys, Joslyn
(1973) greatly increased the aggressiveness of females by
injecting them with androgen and found that among both sexes,
playfighting almost disappeared, "apparently because the males
were simply too frightened by the females to engage in such
activities" (p. 140). Playfighting and aggression have different proximate causes and, in fact, aggression seems to inhibit play more than any other social behavior.

C. *Determinants of Playfighting Success and Rank Among*
    Rhesus Monkeys

Field observation and analysis of motion picture film indicate that during playfights, rhesus monkeys attempt to bite
without being bitten, a goal achieved in part by gaining positions on top of or behind the partner (Symons, 1978). The
attainment of favorable positions is highly nonrandom: during
within-sex playfights, larger animals generally succeed
against smaller animals and, almost independent of size, males
generally succeed against females (Symons, 1978). That
playfighting is competitive might be considered to constitute
evidence in favor of the dominance hypothesis; it might be
argued that playfighting is designed to provide the players
with a low cost answer to the question: if we actually were
fighting, which of us would win? Before this conclusion is
accepted, however, it is worthwhile to consider what is known
about the establishment of rank among rhesus monkeys.

Long term studies of free ranging rhesus monkeys indicate
that the ranks of females and immature males are determined by
the political milieu into which they are born. In large
(Missakian, 1972) and small (Sade, 1967, 1972) groups, adults
are ranked in a linear hierarchy, defined by the direction of
agonistic signals. Within each age group, a juvenile's rank
corresponds to its mother's rank in the adult female hierarchy; as a female matures, she "defeats" the adults subordinate to her mother and comes to rank just below her mother in
the adult hierarchy. A male's eventual rank is less predictable since, beginning at puberty, the male hierarchy becomes
unstable, largely as a result of fighting and changing groups

(See Symons, 1978). (Very similar findings are reported
among Japanese macaques: Kawai, 1965; Kawamura, 1965.)

That genealogy predicts rank seems to be because mothers
support their offspring in agonistic interactions with the
offspring of lower ranking females. By removing and reintro-
ducing animals, Marsden (1968) altered rank order relations in
a captive rhesus group, and found that offspring rose or fell
in rank as their mothers did. In a free-ranging group, a female, orphaned during infancy, eventually attained her genea-
logically-predicted position in the adult hierarchy, suggesting "that dominance relations are formed very early in life
and persist" (Sade, 1967:106). Sade's hypothesis is supported
by Loy and Loy's (1974) study of one and two year old rhesus
that were removed from a single social group and placed together in a corral: ranks remained stable despite the absence
of the juveniles' mothers.

Available data indicate a lack of correspondence between
rhesus playfighting ability and rank. Loy and Loy (1974)
found that yearlings generally dominated two year olds belong-
ing to genealogies more than one rank below their own genealogy, although a yearling is unlikely to outplayfight a two
year old (Symons, 1978). Immature males are generally
much better playfighters than immature females are, and yet in
free ranging rhesus groups, the rank of prepubescent monkeys
is a product of genealogy, not gender (Missakian, 1972; Sade,
1967, 1972).

Changes in rank appear not to be caused by play. The unpredictable rank changes among rhesus males at puberty occur
at a time of increasing aggressiveness -- the consequence of
increasing testosterone production -- and waning playfulness (Symons, 1978). For a time, females dominate their
younger sisters, but a predictable rank reversal occurs, and
they eventually rank in inverse order of age. While a female
may rise above her elder sister as early as one year of age
(Sade, 1967), the reversal usually occurs when the younger
sister is three or four years old (Missakian, 1972), an age
when she plays very infrequently and her elder sisters virtually never play (Symons, 1978). Whatever the cause
such a dominance reversal, its existence contradicts the hypothesis that playfighting determines rank: because of the
difference in body size, a female will have been consistently
inferior to her elder sister in playfighting ability if, indeed, they were close enough in age to have playfought at all.
A male is subordinate to his elder sisters until he is three
years old, at which time he rises above them (Missakian,
1972), but again, this reversal probably is the result of the
increasing aggressiveness that accompanies puberty in males
and not the result of playfighting, since the female will by
this time have ceased or almost ceased to play; furthermore,

if his sister is only a year older than he is, the male may have been the superior playfighter for a year or two before rank reversal.

When rhesus are raised in laboratory groups without their mothers, immature males do outrank immature females (Harlow *et al.*, 1966; Rosenblum, 1961), but this is probably the outcome of greater male aggressiveness (Harlow *et al.*, 1966; Rosenblum, 1961). In Joslyn's (1973) experiment, the androgen injected females replaced males in the top positions in the hierarchy as a result of increased aggressiveness, not increased playfighting abilities, since, as noted above, playfighting was almost nonexistent following the androgen injection.

Data on some canids do not support the view that playfighting establishes rank. Coyotes rarely play until one or more serious fights, between the third and fifth week of life, establish a rank order; wolves and dogs, on the other hand, begin to play long before they establish a rank order, but during this early play period, they are markedly unaggressive (Bekoff, 1977). Playfighting seems to occur primarily ily among animals with established ranks (like rhesus monkeys and coyotes) or among animals that are not in the process of contesting rank (like young dogs and wolves). Where rank is determined by fighting, a correlation between rank and playfighting ability is to be expected, since fighting and playfighting abilities probably are determined by many of the same characteristics, but this is not evidence that the outcomes of playfights produce rank orders. Furthermore, while dominance hierarchies are common throughout the animal kingdom (Wilson, 1975), only a few species play. This observation does not refute the dominance hypothesis, but certainly it must be considered negative evidence.

Although rank does not appear to be determined by the outcomes of playfights, it is my impression that among free-ranging rhesus, play is used occasionally by high ranking animals to bully or intimidate lower ranking animals (Symons, 1978). Both in my study of rhesus play (Symons, 1978) and in Owens' (1975a) study of baboon play, the juvenile daughter of the group's alpha female played very infrequently. In the group that I studied, this yearling female was the second ranking juvenile (her elder brother was first ranking), and she seemed to seek out opportunities to threaten and intimidate other juveniles, especially when her mother was nearby. She playfought most frequently with her elder and younger brothers; other monkeys almost never initiated play with her and, in fact, seemed in subtle ways to avoid her attempts to initiate playfights. Bekoff (1977) reports a strong inverse relation between rank and the ability to successfully solicit play among coyotes; with age, the dominant coyote was

increasingly avoided. One interpretation of these data is
that animals avoid playfighting with those they fear, a practice that seems to make a good deal of sense. A vigorously
playfighting animal is in a vulnerable position should its
partner suddenly attack. The "self-handicapping" that is obvious when play partners are greatly mismatched is an extreme
case of the general inhibition upon which, I believe, the existence of playfighting depends.

D. *Play and Dominance:* *Summary and Conclusions*

While aggression has many proximate causes, its ultimate
cause is competition for scarce resources. From the standpoint of ultimate causation, animals undergo the risks that
aggression entails because, in the long run, successful aggression pays off in reproductive success. Rank orders are
the consequences of compromises made by subordinate animals
who do not contest access to resources or to position because
the probability of success is not great enough to repay the
investment in time, energy, and risk.
    Many lines of evidence indicate that among rhesus monkeys,
aggression and playfighting are incompatible. Among free-
ranging rhesus, dominance among immature animals reflects the
realities of group politics, not their relative playfighting
abilities. Neither do rank reversals appear to result from
reversals in playfighting abilities. Joslyn's (1973) experiment demonstrates that in the absence of mothers, immature
rhesus monkeys rise in rank when they are made more aggressive, but the androgen injected females paid for their high
rank in lost opportunities to playfight. In a free-ranging
group of rhesus and in a free ranging group of baboons, juvenile daughters of the alpha females playfought very little,
perhaps as a consequence of the fear they inspired in their
associates.
    While high rank contributes to reproductive success, an
evolutionary perspective entails the view that frequent, vigorous, sustained, competitive playfighting during youth is
adaptive as well. Perhaps in some cases -- the daughters of
alpha females, for example -- the sacrifice of playfighting
experience is more than offset by the advantages to be
gained from constantly asserting dominance. I found (Symons,
1978) that playfights between males were less likely to become agonistic than were playfights involving a female, perhaps because playfighting experience is more important to male
than to female reproductive success and selection has favored
immature males who inhibit aggression and dominance assertion
during play. Bekoff (in press) reports little evidence in the
literature that animals use play or solicit play in order to

deceive each other. While deception may occur at a level too
subtle to have been detected by observers, an animal that
practiced the crude deception of soliciting play in order to
attack or threaten a subordinate would quickly run out of play
partners. Fear is the foe of play and basis of dominance.

Rank can be established by many means short of a physical
contest: by bluff, by alliance, or by the assessment of the
probable outcome of a fight. Indeed, there probably has been
strong selection for the capacity to assess the competitive
abilities of conspecifics and to calculate risk. But when the
advantages of gaining access to resources or to social position are great enough to repay the risks of a physical contest
and the contestants are matched so closely that neither can
predict the outcome, too much is at stake and too much is at
risk for play-like inhibitions, and the loser of the contest
will flee or submit -- if it can do so without increasing its
peril -- as soon as it detects probable loss. Playfighting is
either a superfluous or an inadequate indicator of who would
win a fight.

The dominance hypothesis of play states in essence that
playfighting functions as a kind of bluff or threat. But if
play had this function, animals could be expected to react to
play initiation or solicitation as they do to a threat, with
submission or, much more rarely, with escalation. Every feature in which playfighting contrasts with fighting -- the inhibition in play, the high frequency of play, the attractive-
ness of play, its regular occurrence between unevenly matched
animals, the initiation of play by subordinate animals and by
obviously outclassed animals, the ability of dominant animals
and animals of superior ability to solicit play successfully,
the failure of playfights to terminate when it becomes clear
which animal is superior -- indicates that playfighting is
not designed to establish rank order and, more to the point,
that the establishment of rank order is unlikely to be an incidental effect or byproduct of playfighting. The adaptive
significance of a playfight is more in the playfighting itself
than in the outcome; the adaptive significance of a status
contest is more in the outcome than in the means by which the
outcome is achieved. Playfights could not establish rank and
remain playful. An activity that is designed appropriately to
establish rank order by physical contest -- in terms of precision, economy, and efficiency -- is fighting. In short, I
believe that the opposite of the dominance hypothesis is correct: frequent, vigorous, sustained, competitive playfighting
can exist only when the players are not thereby contesting
status.

Poole (1973) describes "companion fighting", typical of
familiar male polecats which appears to be intermediate between playfighting and the uninhibited fighting typical of

unfamiliar polecats. In companion fighting, biting is sustained but inhibited, the opponent is not intimidated, and --
unlike uninhibited fighting -- rank order is not established.
Vessey (1968:236) writes that among free ranging rhesus males
"Intergroup play was characterized by its roughness, usually
appearing as something between playing and fighting. Two
animals usually locked up and wrestled vigorously for several
seconds, then separated. Some of the encounters resulted in
chases, but no actual combat ensued from these interactions.
Intragroup play differed in that movements were slower and
there was more contact between participants." I do not think
that I could reliably distinguish intergroup play from some of
the vigorous intragroup play that characterizes males of this
age, but even if intra- and intergroup play should prove to
differ consistently in intensity, I predict that -- as with
companion fighting among male polecats -- the outcomes of intergroup playfights will be found not to establish rank order.

Play apparently occurs in what Reynolds (1976:621) calls
the simulative mode: "The simulative mode of action is para-
doxical: the system's operations should have their normal
consequences, yet those consequences must at the same time be
rendered inconsequential." In deciding what characteristics
of play are compatible with, or are predicted by, functional
hypotheses, to elude Reynolds' paradox, the distinction between ultimate and proximate causation must be maintained:
if the distinction is blurred, inaccurate predictions about
play's immediate causes and motives can be derived from a
valid functional hypothesis, and invalid functional hypotheses
can be generated from an accurate appraisal of play's immediate causes and motives. For example, Leresche (1976:203),
having demonstrated the structural similarities of playfighting and fighting among captive Hamadryas baboons, concludes:
"Instead of viewing play as randomly ordered adult behavior
divorced from its 'real' motivation, we may look at play as
part of the logical process of behavior ontogeny, having an
internal order of its own." But if the argument presented in
this essay is valid, to fulfill its function, whatever that
may be -- in fact, to exist at all -- playfighting must be
divorced from the hostile motives of fighting.

Konner (1975:105) writes: "The extremes of aggressiveness
and competition among males observed in [a] seal species...
may depend on the pattern of socialization in very large peer
play groups of unrelated individuals, in which selection can
be expected to allow high levels of aggressive interaction."
But if playfighting functions to practice and thereby perfect

fighting skills,[7] extremes of adult aggressiveness might, in fact, be promoted more effectively by playfights among relatives than among nonrelatives: assuming, as Konner does, that selection favors higher thresholds of aggression among kin than among nonkin, playfighting among related animals could approximate fighting more closely, and for longer periods of time, without triggering aggression than could playfighting among unrelated animals. Admittedly, this is speculation; how selection actually operates might depend, for example, on whether immature animals are capable of inflicting injury. But the point I wish to emphasize is that playfighting and aggression should not be equated merely because playfighting simulates -- and perhaps functions to perfect -- fighting. In immediate causation, in motivation, and in function, fighting probably has more in common with mild stare threat than it does with playfighting.

Considerations to this point can be summarized thus: function is difficult to demonstrate scientifically; the equation of function and beneficial effect and the influence of functionalist social science compound the difficulties, and may account in part for the popularity of the hypothesis that play functions to establish a dominance order. This hypothesis is not supported by available data and is, in my opinion, improbable, since the foe of play and the basis of dominance is

[7]*Criteria have not been developed for determining conclusively whether playfighting is designed appropriately to function as fighting practice. Intuitive criteria may be inadequate, and the practice hypothesis -- which I find compelling -- frequently is accepted or rejected on insufficient grounds (Symons), 1978). Konner (1975), a supporter of this hypothesis, assumes that an animal invariably derives greater benefit from playfighting with slightly older animals than it derives from playfighting with peers. If Konner is correct, patterns of play initiation should be highly asymmetric; that the predicted asymmetry is absent among rhesus monkeys (Symons, 1978) might, therefore, be counted as evidence that play is not practice. But play with peers and younger animals may, in fact, provide certain kinds of practice benefits not provided by play with older animals. As analogies between animal play and human sports can sometimes provide useful insights (Symons, 1978), it may be of interest that a former national handball champion preferred to practice with good, but not great, players: he maintained that players of his own caliber merely tired him out. Perhaps for each animal there is an ideal profile of play partners, a profile that varies with age, sex, and proficiency.*

fear. The underlying cause of these difficulties, however, may be a folk tradition in which a moral order is manifest in nature.

IV. THE POLITICS OF FUNCTION

As the study of function in biology and the social sciences purports to identify the ultimate causes of the phenomena of life, it is not surprising that this study is widely enmeshed in political controversy. The recent publication of *Sociobiology* (Wilson, 1975) has given this debate new vigor and focus (see, e.g., Allen *et al.*, 1975; Gurin, 1976; Wade, 1976). Critics of sociobiology argue that, regardless of personal beliefs and motives of the sociobiologists themselves, such doctrines inevitably have reactionary political effects. There are two parts to this claim. The first, which I will not consider further, is that emphasizing genetic determinants of behavior makes the status quo appear to be immutable and social reform impossible. The second, articulated most clearly by Sahlins (1976), is that evolutionary biology is the scientific arm of industrial capitalism. Beginning with the "Hobbesian myth", Sahlins argues, social and natural theory have interacted to crystallize and to refine a selfish, individualistic view of humans and nonhuman animals; theories that explain behavior as the outcome of competition, self interest, utilitarianism, or the pursuit of pleasure, justify the inequities of industrial capitalism by making these inequities appear to represent, not the consequences of an arbitrary human institution, but the inevitable working out of a natural process.[8] Marx and Engels enthusiastically supported the idea of evolution and its materialistic implications, believing that their analysis of human society paralleled Darwin's analysis in biology, but they rejected Darwin's major discovery, natural selection, in favor of a Lamarckian mechanism. Venable (1966:64) summarizes Engels' views: "In short, far from supporting Marxism, this theory [natural selection] merely serves, if transferred back from natural

[8]*It is ironic that sociobiologists, who have emphasized kin selection (Hamilton, 1964) and a continuity between kin and group selection (Wilson, 1975), which reduce the importance of individual self interest, should be the primary targets of this charge, when the views of biologists like Ghiselin (1974) and social scientists like McKenzie and Tullock (1975) and Erasmus (1977) are more genuinely individualistic.*

history into the society from which it was originally borrowed, to eternalize and justify as though grounded in nature itself, the barbarous economic relations of the particular historical epoch of bourgeois capitalism."

Paradoxically, while the biological view of organisms as self-interested competitors has been said to have reactionary political implications, in the social sciences the doctrine that societies are integrated, harmonious systems has been judged reactionary: functionalism, it has been argued, justifies inequities by promoting the view that they exist for the benefit of the social system or for the larger good (Martindale, 1965b). Perhaps because primatologists are influenced by functionalism and yet write about matters remote from human politics, the literature on nonhuman primate behavior is replete with examples of functionalism's conservative potential. Consider the following statements about aggression and dominance in the primate behavior textbooks:

Dominance will then probably favor the adults and sacrifice juveniles and infants. This seems adaptive, since the experienced and reproductively active adult is more valuable to the group than an easily replaceable youngster (Kummer, 1971:59).

Primate status roles do not exist except in the social group and can usually be seen as benefiting the group, not the unbridled aggressor (Jolly, 1972:194).

...any social group encounters situations in which some commodity is desired by more animals than the limited amount can supply. A single estrous female cannot simultaneously copulate with two males. Even if she copulates with them in succession, the question of who goes first must be settled. Dominance supplies the answer, but it must be communicated (Simonds, 1974:144).

In the long run, then, dominance promotes tranquility even though it may be based on force (Lancaster, 1975:14).

The very term aggression carries with it an undesirable connotation. Yet if one looks at status hierarchies -- that is, the existence of predictable winners and losers in a society -- as a framework for efficient conflict resolution, these hierarchies have much to recommend them (Bramblett, 1976:96-97).

If, instead of referring to nonhuman primates, these statements had been made about economic and political arrangements among humans in the contemporary United States, I suspect that they would have a decidedly conservative ring.

Referring to the emphases on hierarchy and territory in the literature on nonhuman primate behavior, Rowell (1972:159) writes: "It is surely a comment on our own species that we have attempted to explain the behaviour of other species almost entirely in terms of concepts defined by aggression." But, as the above textbook quotations illustrate, the concept of hierarchy is often used not so much to explain behavior in terms of aggression as it is to explain aggression away, to show that apparently self serving behavior, when looked at correctly, actually is designed to benefit the commonweal. In primatology, the word "dominance" has become a euphemism for aggression. Similarly, primatologists regularly discuss the need for defense among nonhuman primates, and the defensive role played by certain members of the group, but not the need for offense. It seems likely that this imbalance arises from the same motive that leads nations to spend billions for defense but not one cent for offense: to put actions in the best possible light. The result of these trends in primatology is not, as Rowell implies, an overemphasis on the importance of aggression; on the contrary, the importance of intraspecific aggression as a cause of mortality has been greatly underestimated (Hamburg and Goodall, in press; Marler, 1976; Symons, 1978).

In *The Selfish Gene*, Dawkins (1976:10) writes: "Perhaps one reason for the great appeal of the group-selection theory is that it is thoroughly in tune with the moral and political ideals which most of us share." But a scientific theory can be in or out of tune with morality only from the perspective of a folk cosmology: "it is most unusual to find a people before modern times who did not find principles of human conduct and organization in nature, either immanent or put there by divinities" (Sivin, 1973:xxvii). Sivin also remarks that "today we do not consider nature an ethical order, nor human society a little cosmos", but this may be too optimistic. The idea of a unity and harmony in nature which is a model for the conduct of human affairs is at least as old as Western history. Sumerian theologians believed the universe to be superintended by manlike, but superhuman and immortal, beings who govern according to established rules and regulations, and "To the Ionian philosophers, the cosmos meant an arrangement of all things in which every natural power has its function and its limits assigned. As in any good arrangement, this implies a systematic unity in which diverse elements are combined or composed" (Glacken, 1967:16-17). In Judeo-Christian tradition, the good is defined by the will of God and is assumed to be manifest in nature. Although natural creative processes have largely replaced supernatural entity, and God no longer figures prominently in the natural and social sciences, the underlying belief in a benevolent order in nature often persists, as does the tendency to seek the good in whatever processes are believed to create organic and social design. According to Williams (1966:254), a great deal of misunderstanding in biology has resulted from attempts "to find not only an order in nature but a moral order." The similarities of group selection theory in biology and functionalism in the social sciences may reflect their common origin in a view of the universe as the peaceable kingdom.[9] Perhaps one reason that function is so regularly equated with beneficial effect is that this makes it possible to emphasize effects that give nature the appearance of a moral order and to deemphasize natural selection's amorality.

The notion that science can confirm or refute moral precepts is bad philosophy. In *Treatise of Human Nature*, David Hume demonstrated the error in deducing normative conclusions from descriptive premises: "is" and "is not" specify fundamentally different kinds of relations than are specified by "ought" and "ought not" (see Flew, 1967). Whether there are systematic relationships between beliefs about human nature and its ultimate causes, on the one hand, and other beliefs and behaviors, on the other, is a matter for the social sciences and for social, political, and intellectual history; if such relationships exist, they are empirical, and are not relations of logical necessity. Functionalism does not logically entail political conservatism, nor does Darwinism logically entail Social Darwinism.

But given the tendency in folk tradition to find good in whatever processes are believed to create organic and social forms, it is not surprising that scientists continue to appeal to natural selection or to sociocultural selection to justify moral statements. Sperry (1977:243), for example, writes that "what is good, right, or to be valued is defined very broadly to be that which accords with, sustains, and enhances the orderly design of evolving nature. Conversely,

[9]*"Group selection" has come to stand for altruism and a tender-minded view of life, individual/kin selection for selfishness and a tough-minded view of life. This dichotomy is questionable. The ordinary meanings of "altruism" and "selfishness" have little to do with their meanings in biology, where they refer only to the survival of genes. Group selection might promote the evolution of willing cannon fodder and, hence, a nature redder in tooth and claw than exists in reality or the Victorian imagination [see, e.g., Maynard Smith's (1972) discussion of the role of "altruism" in making insect "war" possible].*

Donald Symons

whatever is out of line, degrades, or destroys nature's grand
design is wrong or bad."[10] Sperry's grand design is reminis-
cent of Western cosmologies, but even scholars with less san-
guine views of nature occasionally make similar pronouncements.
DeVore (1977), for example, after explicitly denying that
natural selection is to be identified with the good, nevertheless goes on to say that his scientific view -- that human behaviors are facultative adaptive strategies, molded by natural
selection to maximize reproductive success at the expense of
other humans -- "ultimately lends a certain dignity to behaviors that one might otherwise consider aberrant or animalis-
tic" (p. 87). On the other hand, Campbell (1975), whose views
on human nature and the selective process that produced it are
the same as DeVore's, does not find dignity in natural selection; he maintains that sociocultural selection counters natural, selfish human impulses (ignoring the possibility that institutions, values, laws, religions, and moral traditions represent the cumulative outcomes of "selfish" individuals pur-
suing their own advantage). Campbell thus concludes that
sociocultural selection is to be identified with the good, and
hence that "social systems" are invested with wisdom and --
merely because they exist -- should be viewed with awe and respect. (Critical discussions of Campbell's paper can be found
in the *American Psychologist 31*:341-384.)

Many scholars believe that although science cannot determine what is right, it can predict what social arrangements
are likely to be workable (Klein, personal communication).
Yet Sivin (1973:xxix) writes: "Modern science has nothing to
contribute to moral reflection except predictions of consequences. Society at large retains little confidence even in
the objectivity of these predictions, since in any issue of
public moment it is almost inevitable that both sides will
produce duly accredited scientists to predict what are represented as diametrically opposite outcomes."

The belief that a moral order is to be found in nature is
not just bad philosophy, it is exceedingly dangerous, because
it renders moral and ethical precepts vulnerable to scientific
disproof. What if our ethical systems are not reflected in
nature? What if it should turn out that "The economy of nature is competitive from beginning to end" (Ghiselin, 1974);

[10]*Sperry attempts to insulate his argument from philosophical analysis by maintaining, in effect, that the question of
whether "ought" can be derived from "is" is itself a scientific question; he concludes that recent scientific developments justify such a derivation, and dismisses philosophical
analysis as "a logical artifact of a strictly pencil-and-paper
approach" (p. 241).*

that "'nature red in tooth and claw' sums up our modern understanding of natural selection admirably" (Dawkins, 1976); that
animal playfights are designed to develop, not creativity and
responsible citizenship, but skill in violence? Lack (1969)
points out that the world of birds often strikes us as idyl-
lic, a model for human societies to emulate, but that this
view of birds results from anthropomorphism and ignorance:
"We would not enjoy a society in which one-third of our adult
friends and over four-fifths of the teenagers die of starvation each year" (p. 21).

When the chimpanzee was thought to be an unaggressive,
promiscuous, Rousseauian ape, it was often endorsed in preference to the unpleasant, dominance seeking baboon as a representative early hominid. Now that recent field studies of
chimpanzees have revealed sequestering of estrous females, extreme territoriality with border patrolling, deliberate attempts by groups to kill helpless, lone conspecifics, and the
consumption of living infants (Hamburg and Goodall, in press),
those who implied that our view of human nature depends on
which nonhuman primate we choose as a model ancestor may find
it difficult to backtrack. Similarly, the attempt to make
nonhuman primates cozy and familiar by using words like "aunt"
backfires in the specter of "aunting to death" (Hrdy, 1976).

Thomas Huxley (1897) pointed out that from a wholly intellectual point of view, nature appears to be beautiful and harmonious, but if we allow moral sympathies to influence our
judgment, our view of nature is a darker one: "In sober
truth, to those who have made a study of the phenomena of life
as they are exhibited by the higher forms of the animal world,
the optimistic dogma, that this is the best of all possible
worlds, will seem little better than libel upon possibility"
(p. 196). Huxley contrasted the state of nature, and the cosmic process (evolution by natural selection) of which it is
the outcome, with the state of art, produced by human intelligence, exemplified by a garden. The garden can be maintained
only by counteracting the forces of nature, creating an artificial environment in which Malthusian reproductive competition is restrained and, hence, the struggle for existence is
largely arrested. Like the garden, the kind of society in
which most of us would like to live can exist only by virtue
of "artificial" ethics, in opposition to the cosmic process,
since "with all their enormous differences in natural endow-
ment, men agree in one thing, and that is their innate desire
to enjoy the pleasures and to escape the pains of life; and,
in short, to do nothing but that which it pleases them to do,
without the least reference to the welfare of the society into
which they are born. That is their inheritance (the reality
at the bottom of the doctrine of original sin..." (p. 27).

Freud wrote in *Civilization and Its Discontents:* "men are

not gentle creatures who want to be loved, and who at the most
can defend themselves if they are attacked...their neighbour
is for them not only a potential helper or sexual object, but
also someone who tempts them to satisfy their aggressiveness
on him, to exploit his capacity for work without compensation,
to use him sexually without his consent, to seize his possessions, to humiliate him, to cause him pain, to torture and to
kill him." If group selection were found to play an important
role in the evolutionary process, and biologists announced
that "altruism" was now natural, the historical record to
which Freud referred would remain unaltered. This record, not
the pronouncements of evolutionary biologists, is the real
source of evidence about human behavior and human nature.

V. SUMMARY

    In attempting to clarify the question of function in prima-
tology, fundamentally different kinds of statements have been
made in the course of this essay, and these statements have
been supported by different kinds of evidence. First, scientific statements, primarily intended to demonstrate that, contrary to most primatological thinking, play is unlikely to establish a dominance order, were supported by the usual kinds
of theoretical and factual data and citations. Furthermore,
it was argued that in primatology, many functional hypotheses
have little basis in present understandings of animal behavior
and evolutionary theory.
    Second, statements were made about the history of prima-
tology. It was suggested that social science concepts of
function, and the attempt to find a moral order in nature,
have impeded the study of function in primatology and may be
responsible for much fruitless functional speculation. These
statements were supported largely by circumstantial evidence
and are susceptible to more rigorous treatment.
    Third, statements were made about how primatology ought to
be conducted to promote progress. It was argued that the
study of the purposes for which behaviors have been designed
by natural selection is a useful, albeit difficult and primitive, scientific discipline, whereas the study of the beneficial effects of behavior -- which in primatology has often
been equated with the study of function -- is not a useful
discipline any more than would be a branch of biology devoted
to the study of organisms weighing more than 1.7 kilograms.
These statements were supported by appeal to the arguments of
evolutionary biologists such as Williams and Ghiselin and students of animal behavior such as Tinbergen and Hinde.
    Fourth, a philosophical position was taken: what ought to

be cannot be derived from what is. This position was supported by appeal to a philosopher, David Hume, and echoes mainstream philosophical thinking. The study of function in primate behavior might be promoted if primatology in general were to adopt this position, since nonhuman animals would thereby cease to be moral exemplars.

ACKNOWLEDGMENTS

I thank the following individuals who read and criticized earlier drafts of this essay: Owen Aldis, Marc Bekoff, Naomi Bishop, D. E. Brown, Ivan Chase, Charles Erasmus, Robert Fagen, Elvin Hatch, Lewis Klein, Jane Lancaster, Donald Lindburg, Priscilla Robertson, Euclid Smith, John Townsend, Barbara Voorhies, Phillip Walker, and Sherwood Washburn. The work benefitted substantially from the advice of these scholars; as I did not incorporate every suggestion, however, I, alone, am responsible for deficiencies that remain. Charlotte Symons edited the manuscript, for which I am most grateful. Finally, I thank Lewis Klein, who clearly delineated the issues underlying the group selection controversy more than a decade ago.

REFERENCES

Aldis, O. "Play Fighting". Academic Press, New York (1975).
Allen, E., Beckwith, B., Beckwith, J., Chorover, S., Culver, D., Duncan, M., Gould, S., Hubbard, R., Inouye, H., Leeds, A., Lewontin, R., Madansky, C., Miller, L., Pyeritz, R., Rosenthal, M., and Schreier, H. Against sociobiology. *The New York Review of Books*, November 13: 43-44 (1975).
Baldwin, J. D., and Baldwin, J. I. The role of play in social organization: comparative observations on squirrel monkeys (*Saimiri*). *Primates 14*, 369-381 (1973).
_____ Exploration and social play in squirrel monkeys (*Saimiri*). *Amer. Zool. 14*, 303-315 (1974).
_____ Effects of food ecology on social play: a laboratory simulation. *Z. Tierpsychol. 40*, 1-14 (1976).
_____ The role of learning phenomena in the ontogeny of exploration and play, *in* "Primate Bio-Social Development: Biological, Social, and Ecological Determinants" (S. Chevalier-Skolnikoff, and F. E. Poirier, eds.), pp. 343-406. Garland Publishing Co., New York (1977).
Barth, F. "Models of Social Organization". Royal Anthropological Institute of Great Britain and Ireland occasional paper No. 23 (1966).

Beach, F. A.   Current concepts of play in animals.   *Amer. Nat.*
79, 523-541 (1945).

Beer, C. G.   Multiple functions and gull displays, *in* "Function and Evolution in Behaviour" (G. Baerends, C. Beer,
and A. Manning, eds.), pp. 16-54.   Clarendon Press, Oxford (1975).

Bekoff, M.   The social deprivation paradigm: who's being deprived of what?   *Develop. Psychobiol. 9*, 499-500 (1976).

_____   Mammalian dispersal and the ontogeny of individual behavioral phenotypes.   *Amer. Nat. 111*, 715-732 (1977).

_____   Social play: structure, function, and the evolution of
a cooperative social behavior, *in* "Comparative and Evolutionary Aspects of Behavioral Development" (G. M. Burg-
hardt, and M. Bekoff, eds.), Garland Publishing Co., New
York (in press).

Boissevain, J.   "Friends of Friends".   Basil Blackwell, Oxford
(1974).

Bramblett, C. A.   "Patterns of Primate Behavior".   Mayfield,
Palo Alto, California (1976).

Campbell, D. T.   On the conflicts between biological and social evolution and between psychology and moral tradition.
Amer. Psychol. 30, 1103-1126 (1975).

Carpenter, C. R.   A field study of the behavior and social relations of howling monkeys.   *Comp. Psychol. Monogr. 10*(2)
(1934).

Cartmill, M.   Rethinking primate origins.   *Science 184,* 436-
443 (1974).

Cloak, F. T., Jr.   Is a cultural ethology possible?   *Human
Ecology 3*, 161-182 (1975).

Clutton-Brock, T. H., and Harvey, P. H.   Evolutionary rules
and primate societies, *in* "Growing Points in Ethology"
(P. P. G. Bateson, and R. A. Hinde, eds.), pp. 195-237.
Cambridge University Press, Cambridge (1976).

Dawkins, R.   "The Selfish Gene".   Oxford University Press,
Oxford (1976).

Delgado, J. M. R.   Communication with the conscious brain by
means of electrical and chemical probes, *in* "Factors in
Depression" (N. S. Kline, ed.), pp. 251-268.   Raven Press,
New York (1974).

DeVore, I.   The new science of genetic self-interest (inter-
view with I. DeVore by S. Morris).   *Psychology Today,*
February, 42 (1977).

Dolhinow, P. J.   At play in the fields.   *Natural History 80,*
66-71 (1971).

Dolhinow, P. J., and Bishop, N.   The development of motor
skills and social relationships among primates through
play.   *Minnesota Symposia on Child Psychology 4,* 141-198
(1970).

Erasmus, C. J.   "In Search of the Common Good: Utopian

Experiments Past and Future". The Free Press, New York (1977).

Fagen, R. Selective and evolutionary aspects of animal play. *Amer. Nat. 108,* 850-858 (1974).

_____ Exercise, play, and physical training in animals, *in* "Perspectives in Ethology, Vol. 2" (P. P. G. Bateson, and P. H. Klopfer, eds.), pp. 189-219. Plenum Press, New York (1976).

Flew, A. "Evolutionary Ethics". St. Martin's Press, New York (1967).

Ghiselin, M. T. "The Economy of Nature and the Evolution of Sex". The University of California Press, Berkeley (1974).

Glacken, C. J. "Traces on the Rhodian Shore". The University of California Press, Berkeley (1967).

Gottier, R. F. Factors affecting agonistic behavior in several subhuman species. *Genetic Psychol. Monogr. 86,* 177-218 (1972).

Gurin, J. Is society hereditary? *Harvard Magazine 79,* 21-25 (1976).

Hall, K. R. L. Behaviour and ecology of the wild patas monkey, *Erythrocebus patas,* in Uganda. *J. Zool. 148,* 15-87 (1965).

Hamburg, D., and Goodall, J. (eds.). "Perspectives on Human Evolution, Vol. 5". W. A. Benjamin, Menlo Park, California (in press).

Hamilton, W. D. The genetical evolution of social behavior. *J. theoret. Biol. 7,* 1-52 (1964).

Harlow, H. F. Age mate or peer affectional system, *in* "Advances in the Study of Behaviour, Vol. 2" (D. S. Lehrman, R. A. Hinde, and E. Shore, eds.), pp. 333-383. Academic Press, New York (1969).

Harlow, H. F., and Harlow, M. K. The affectional systems, *in* "Behavior of Nonhuman Primates, Vol. 2" (A. M. Schrier, H. F. Harlow, and F. Stollnitz, eds.), pp. 287-334. Academic Press, New York (1965).

Harlow, H. F., Joslyn, W. D., Senko, M. G., and Dopp, A. Behavioral aspects of reproduction in primates. *J. Anim. Sci. 25,* 49-67 (1966).

Hatch, E. "Theories of Man and Culture". Columbia University Press, New York (1973).

Hasufater, G. "Dominance and Reproduction in Baboons *(Papio cynocephalus)*: A Quantitative Analysis". *Contrib. Primatol. 7,* 1-148 (1976).

Hempel, C. G. The logic of functional analysis, *in* "Symposium on Sociological Theory" (L. Gross, ed.), pp. 271-307. Harper & Row, New York (1959).

Hinde, R. A. The concept of function, *in* "Function and Evolution in Behaviour" (F. Baerends, C. Beer, and A. Manning, eds.), pp. 3-15. Oxford University Press, Oxford (1975).

_____ The nature of social structure, *in* "Perspectives on

Human Evolution, Vol. 5" (D. Hamburg, and J. Goodall,
    eds.). W. A. Benjamin, Menlo Park, California (in press).
Holt, R. T. A proposed structural-functional framework for
    political science, *in* "Functionalism in the Social Scien-
    ces: The Strength and Limits of Functionalism in Anthro-
    pology, Economics, Political Science, and Sociology" (D.
    Martindale, ed.), pp. 84-110. The American Academy of
    Political and Social Science, Philadelphia (1965).
Hrdy, S. B. Care and exploitation of nonhuman primate infants
    by conspecifics other than the mother, *in* "Advances in the
    Study of Behavior, Vol. 6", pp. 101-158. Academic Press,
    New York (1976).
Hull, D. L. "Philosophy of Biological Science". Prentice-
    Hall, Englewood Cliffs, New Jersey (1974).
Huxley, T. H. "Evolution and Ethics and Other Essays". D.
    Appleton, New York (1897).
Jarvie, I. C. "The Revolution in Anthropology". The Humani-
    ties Press, New York (1964).
_____ Limits of functionalism and alternatives to it in an-
    thropology, *in* "Functionalism in the Social Sciences: The
    Strength and Limits of Functionalism in Anthropology,
    Economics, Political Science, and Sociology" (D. Martin-
    dale, ed.), pp. 18-34. The American Academy of Political
    and Social Science, Philadelphia (1965).
_____ "Functionalism". Burgess, Minneapolis (1973).
Jolly, A. "The Evolution of Primate Behavior". Macmillan
    Publishing Co., New York (1972).
Joslyn, W. D. Androgen-induced social dominance in infant
    female rhesus monkeys. *J. Child Psychol. Psychiat. 14*,
    137-145 (1973).
Kawai, M. On the system of social ranks in a natural troop of
    Japanese monkeys: (I) basic rank and dependent rank, *in*
    "Japanese Monkeys: A Collection of Translations" (S. A.
    Altmann, ed.), pp. 66-86. Published by the editor, Al-
    berta (1965).
Kawamura, S. Matriarchal social ranks in the Minoo-B troop: a
    study of the rank system of Japanese monkeys, *in* "Japanese
    Monkeys: A Collection of Translations" (S. A. Altmann,
    ed.), pp. 105-112. Published by the editor, Alberta (1965).
Klein, L. Group selection and food ecology. Unpublished man-
    uscript (1966).
Konner, M. Relations among infants and juveniles in compara-
    tive perspective, *in* "Friendship and Peer Relations" (M.
    Lewis, and L. A. Rosenblum, ed.), pp. 99-129. John Wiley
    and Sons, New York (1975).
Krupp, S. R. Equilibrium theory in economics and in func-
    tional analysis as types of explanation, *in* "Functionalism
    in the Social Sciences: The Strength and Limits of Func-
    tionalism in Anthropology, Economics, Political Science,

and Sociology" (D. Martindale, ed.), pp. 65-83. The American Academy of Political and Social Science, Philadelphia (1965).

Kummer, H. "Primate Societies". Aldine, Chicago (1971).

Lack, D. "Population Studies of Birds". Oxford University Press, Oxford (1966).

_____ Of birds and men. *New Scientist 16,* 121-122 (1969).

Lancaster, J. B. "Primate Behavior and the Emergence of Human Culture". Holt, Rinehart and Winston, New York (1975).

Leach, E. R. "Pul Eliya: A Village in Ceylon". Cambridge University Press, Cambridge (1961).

Lehrman, D. S. Semantic and conceptual issues in the nature-nurture problem, *in* "Development and Evolution of Behavior" (L. R. Aronson, and E. Tobach, eds.), pp. 17-52. W. H. Freeman, New York (1970).

Leresche, L. A. Dyadic play in hamadryas baboons. *Behaviour 57,* 190-205 (1976).

Loy, J., and Loy, K. Behavior of an all-juvenile group of rhesus monkeys. *Am. J. phys. Anthrop. 40,* 83-96 (1974).

Malinowski, B. "Sex and Repression in Savage Society". Routledge & Kegan Paul, London (1927).

_____ The group and the individual in functional analysis. *Amer. J. Sociol. 44,* 938-964 (1939).

Marler, P. On animal aggression. *Amer. Psychol. 31,* 239-246 (1976).

Marsden, H. M. Agonistic behaviour of young rhesus monkeys after changes induced in social rank of their mothers. *Anim. Behav. 16,* 38-44 (1968).

Martindale, D. Foreword, *in* "Functionalism in the Social Sciences: The Strength and Limits of Functionalism in Anthropology, Economics, Political Science, and Sociology" (D. Martindale, ed.), pp. vii-ix. The American Academy of Political and Social Science, Philadelphia (1965a).

_____ Limits of and alternatives to functionalism in sociology, *in* "Functionalism in the Social Sciences: The Strength and Limits of Functionalism in Anthropology, Economics, Political Science, and Sociology" (D. Martindale, ed.), pp. 144-162. The American Academy of Political and Social Science, Philadelphia (1965b).

Maynard Smith, J. Game theory and the evolution of fighting, *in* "On Evolution" (J. Maynard Smith, ed.), pp. 8-28. Edinburgh University Press, Edinburgh (1972).

_____ The theory of games and the evolution of animal conflicts. *J. theor. Biol. 47,* 209-221 (1974).

_____ Group selection. *Quart. Rev. Biol. 51,* 277-283 (1976).

Maynard Smith, J., and Price, G. R. The logic of animal conflict. *Nature 246,* 15-18 (1973).

Mayr, E. Cause and effect in biology. *Science 134,* 1501-1506 (1961).

McKenzie, R. B., and Tullock, G. "The New World of Economics: Explorations into the Human Experience". Richard D. Irwin, Homewood, Illinois (1975).

Meier, G. W., and Devanney, V. D. The ontogeny of play within a society: preliminary analysis. *Amer. Zool. 14*, 289-294 (1974).

Merton, R. K. "Social Theory and Social Structure". The Free Press, Glencoe, Illinois (1957).

Millar, S. "The Psychology of Play". Penguin Books, Baltimore (1968).

Missakian, E. A. Genealogical and cross-genealogical dominance relations in a group of free-ranging monkeys *(Macaca mulatta)* on Cayo Santiago. *Primates 13*, 169-180 (1972).

Oakley, F. B., and Reynolds, P. C. Differing responses to social play deprivation in two species of macaque, *in* "The Anthropological Study of Play: Problems and Prospects" (D. F. Lancy, and B. A. Tindall, eds.), pp. 179-188. Leisure Press, Cornwall, New York (1976).

Owens, N. W. Social play behaviour in free-living baboons, *Papio anubis. Anim. Behav. 23*, 387-408 (1975a).

_____ A comparison of aggressive play and aggression in free-living baboons, *Papio anubis. Anim. Behav. 23*, 757-765 (1975b).

Parker, G. A. Assessment strategy and the evolution of fighting behaviour. *J. theor. Biol. 47*, 223-243 (1974).

Poirier, F. E. The Nilgiri langur of south India, *in* "Primate Behavior, Vol. 1" (L. A. Rosenblum, ed.), pp. 254-383. Academic Press, New York (1970).

_____ Introduction, *in* "Primate Socialization" (F. E. Poirier, ed.), pp. 3-28. Random House, New York (1972).

Poirier, F. E., and Smith, E. O. Socializing functions of primate play. *Amer. Zool. 14*, 275-287 (1974).

Poole, T. B. The aggressive behaviour of individual male polecats *(Mustela putorius, M. furo* and hybrids*)* towards familiar and unfamiliar opponents. *J. Zool. 170*, 395-414 (1973).

Popp, J. L., and DeVore, I. Aggressive competition and social dominance theory, *in* "Perspectives on Human Evolution, Vol. 5" (D. Hamburg, and J. Goodall, eds.). W. A. Benjamin, Menlo Park, California (in press).

Radcliffe-Brown, A. R. On the concept of function in social science. *Amer. Anthrop. 37*, 394-402 (1935).

Reynolds, P. C. Play, language and human evolution, *in* "Play: Its Role in Development and Evolution" (J. S. Bruner, A. Jolly, and K. Sylva, eds.), pp. 621-635. Basic Books, New York (1976).

Richerson, P. J. Ecology and human ecology: a comparison of theories in the biological and social sciences. *Amer. Ethnol. 4*, 1-26 (1977).

Rosenblum, L. A.   The Development of Social Behavior in the
    Rhesus Monkey.   Ph.D. Dissertation, University of Wiscon-
    sin (1961).
Rowell, T. E.   "The Social Behaviour of Monkeys".   Penguin
    Books, Baltimore (1972).
_____ Growing up in a monkey group.   Ethos 3, 113-128 (1975).
Ruyle, E. E.   Genetic and cultural pools: some suggestions
    for a unified theory of biocultural evolution.   Human
    Ecology 1, 201-215 (1973).
Sade, D. S.   Determinants of dominance in a group of free-
    ranging rhesus monkeys, in "Social Communication among
    Primates" (S. A. Altmann, ed.), pp. 99-114.   University of
    Chicago Press, Chicago (1967).
_____ A longitudinal study of social behavior of rhesus mon-
    keys, in "The Functional and Evolutionary Biology of Pri-
    mates" (R. H. Tuttle, ed.), pp. 378-398.   Aldine, Chicago
    (1972).
Sahlins, M.   "The Use and Abuse of Biology: An Anthropologi-
    cal Critique of Sociobiology".   The University of Michigan
    Press, Ann Arbor (1976).
Simonds, P. E.   "The Social Primates".   Harper & Row, New York
    (1974).
Sivin, N.   Preface, in "Chinese Science: Explorations of an
    Ancient Tradition" (S. Nakayama, and N. Sivin, eds.),
    pp. i-xxxvi.   The MIT Press, Cambridge, Massachusetts
    (1973).
Spencer, R. F.   The nature and value of functionalism in an-
    thropology, in "Functionalism in the Social Sciences:
    The Strength and Limits of Functionalism in Anthropology,
    Economics, Political Science, and Sociology" (D. Martin-
    dale, ed.), pp. 1-17.   The American Academy of Political
    and Social Science, Philadelphia (1965).
Sperry, R. W.   Bridging science and values.   Amer. Psychol.
    32, 237-245 (1977).
Suomi, S. J., and Harlow, H. F.   Monkeys at play.   Natural
    History, 80, 72-75 (1971).
Symons, D.   Aggressive play and communication in rhesus mon-
    keys (Macaca mulatta).   Amer. Zool. 14, 317-322 (1974).
_____ "Play and Aggression: A Study of Rhesus Monkeys".
    Columbia University Press, New York (1978).
Thompson, N.   My descent from the monkey, in "Perspectives in
    Ethology, Vol. 2" (P. P. G. Bateson, and P. H. Klopfer,
    eds.), pp. 221-230.   Plenum Press, New York (1976).
Tinbergen, N.   Behavior and natural selection, in "Ideas in
    Modern Biology" (Proc. XVI Int. Congr. Zool., Vol. 6),
    pp. 519-542.   The Natural History Press, Garden City, New
    York (1965).
_____ Adaptive features of the black-headed gull, in Proc.
    XIV Int. Ornithol. Congr. (D. W. Snow, ed.), pp. 43-59.

Blackwell Scientific Publications, Oxford (1967).

Venable, V. "Human Nature: The Marxian View". The World Publishing Company, Cleveland (1966).

Vessey, S. H. Interactions between free-ranging groups of rhesus monkeys. *Folia primat. 8,* 228-239 (1968).

Wade, N. Sociobiology: troubled birth for new discipline. *Science 191,* 1151-1155 (1976).

Washburn, S. L., and Hamburg, D. A. Aggressive behavior in Old World monkeys and apes, *in* "Primates: Studies in Adaptation and Variability" (P. C. Jay, ed.), pp. 458-478. Holt, Rinehart and Winston, New York (1968).

Williams, G. C. "Adaptation and Natural Selection". Princeton University Press, Princeton (1966).

Wilson, E. O. "Sociobiology: The New Synthesis". The Belknap Press of Harvard University Press, Cambridge, Massachusetts (1975).

REINFORCEMENT THEORIES OF EXPLORATION, PLAY, CREATIVITY AND PSYCHOSOCIAL GROWTH[1]

John D. Baldwin
Janice I. Baldwin

Department of Sociology
University of California at Santa Barbara
Santa Barbara, California

In a previous paper, we attempted to show that in primates, and probably other higher animals, exploration and play are learned under contingencies imposed primarily by sensory stimulation reinforcers (Baldwin and Baldwin, 1977). The purposes of this paper are to i) delineate the basic mechanisms that influence the contingencies of sensory stimulation (SS) reinforcement, ii) trace the ontogeny of exploration and play (e&p) in the developing individual, iii) explain the relation between SS reinforcers and 6 phases of e&p, iv) show how creativity and psychosocial growth in primates (Ps) often result from the contingencies that condition e&p, v) explain how biological mechanisms for e&p dovetail with reinforcement mechanisms, vi) describe 20 experiments that could be conducted to test the present theory, and vii) extend this work to include some of the humanistic implications of the theory for enhancing exploration, play, creativity and psychosocial growth in humans.

Although the symposium's goal is to focus on social play, this chapter will discuss both social and nonsocial forms of exploration, play and creativity. It is our position that the

---

[1] We would like to dedicate this paper to the memory of the late Frank DuMond. Frank's great generosity in making his primate facilities available for scientific studies has helped many primatologists collect data that could not be obtained elsewhere. We, personally, are deeply indebted for his repeated assistance in our studies of the squirrel monkey troop at his New World primate display.

controlling variables responsible for exploration and creativity are very similar to those for play and that social forms of these activities are not qualitatively different from nonsocial forms.

I. THE REINFORCERS FOR EXPLORATION AND PLAY

We will begin with a discussion of the basic, nonsocial reinforcers that control *e&p*, but later extend the analysis to include social learning factors -- social reinforcement, observational learning, and so forth.

Welker (1971:180) provides a useful schema for approaching the categorization and definition of different types of exploration. "Pure" exploration occurs when an animal orients to novel stimuli; and exploration "in the 'service' of other 'needs'" occurs when the animal is motivated by other reinforcers, such as food, sex or escape of pain. We wish to expand on this dichotomy by recognizing that both exploration *and* play can be oriented to novel experience or function in the service of other motivations. Expanding further on Welker's dichotomy, we place pure *e&p* at one end of a continuum and *e&p* in the service of other reinforcers toward the other end (Figure 1). Most normal *e&p* is somewhat influenced by multiple factors, hence lies to the right of pure *e&p*. Moving to mixed *e&p*[2] and then to *e&p* in the service of other motivations, pure *e&p* becomes increasingly dominated by feeding, making sexual contact and obtaining the other reinforcers. At the right end of the continuum, the motivation for *e&p* has been totally replaced by the other motivations, and the animals seek food, sex, warmth, safety and other reinforcers with no signs of *e&p*.

During the development of the individual, there is a tendency for young *P*s to spend more time in pure and normal *e&p*, then to move progressively toward the right end of the continuum with increasing age and experience, as other reinforcers come to dominate. In this paper, we will limit ourselves to discussions of pure or relatively pure types of *e&p* in order to describe the controlling variables in their clearest form. The conditions that produce other forms of *e&p* are discussed elsewhere (Baldwin and Baldwin, 1977).

If we are defining the impure types of *e&p* in terms of their service to other reinforcers, then clearly we must specify the reinforcers for the behavior at the left end of the continuum. Many observers have concluded that *e&p* are moti-

----

[2]*Examples of mixed* e&p *include sex play, play hunting and exploring various foods while partially satiated with food.*

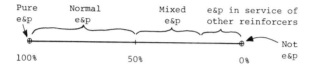

FIGURE 1. *Various types of e&p are shown on a continuum that indicates the degree (0 to 100%) to which SS is the controlling reinforcer.*

vated by various aspects of *SS*: such as stimulus complexity, novelty, stimulus variety, unpredictable experience and stimulus intensity (Berlyne, 1960; Fiske and Maddi, 1961; Hebb, 1966; Humphrey, 1974; Mason, 1968; Simpson, 1976; Welker, 1971). Butler (1958, 1965), Campbell (1972), Ellis (1973), Schultz (1965) and others have demonstrated that *SS* can serve as the reinforcer for *e&p*. It follows that pure *e&p* can be defined as "stimulus-seeking behavior", since stimulus-seeking is the underlying feature of *e&p* that leads to the controlling reinforcers. This definition focuses on the controlling factors that cause *e&p* rather than on the topographies of *e&p* behavior -- which can be enormously complex and varied. The fact that animals will bar press for access to novel flashing lights, sounds, views of outside events or pictures, and so forth, indicates that novel or varied sensory input can serve as a reinforcer. However, extremely high levels of *SS* are overarousing and aversive, and very low levels of *SS* are underarousing, boring, and, hence, also aversive: *SS* is a positive reinforcer only at intermediate levels (Ellis, 1973).

The reason that an intermediate level of *SS* is a positive
reinforcer appears to relate to the optimal arousal levels in
the CNS that result from intermediate stimulus input (Fiske
and Maddi, 1961; Hebb, 1955, 1966). Hebb has described the
basic correlation between levels of behavioral performance and
central arousal as following an inverted "U" pattern (Figure
2a). At low arousal levels, the organism is drowsy, ineffic-
ient, uncoordinated and, hence, poorly prepared to perform
most tasks. As arousal rises into the optimal zone, the organism becomes alert, responsive, more able to perform complex
associations and, hence, more capable of a high level of behavioral performance. When arousal exceeds the optimal zone,
performance begins to degenerate due to increased distractions,
overload of processing systems and stress.

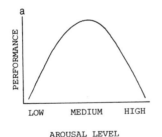
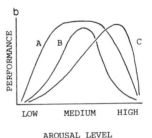

FIGURE 2. a) Performance usually reaches peak levels at
intermediate arousal. b) Overlearned behavior (A) can be performed well over a broader range of arousal levels than underlearned behavior (B); whereas high muscle tension behavior (C)
is best performed at high arousal levels.

Hebb's formulation also takes into consideration the varying demand characteristics of different types of behavior
(Figure 2b). Overlearned behavior (type A) can be performed
well over a wider range of arousal levels than underlearned
behavior (type B); and high muscle tension behavior, such as
fight or flight (type C), is more compatible with high arousal
zones than most other behavior. If the brain has optimal
arousal levels in which performance and survival are maximized,
it is easy to see that natural selection would favor the evolution of mechanisms which keep an organism in its optimal arousal zone. The reticular formation of the brain stem has been
implicated as the homeostat that functions to monitor central
arousal level (Jasper, 1958; Lindsley, 1951, 1961; Samuels,
1959), which correlates with its diffuse structure, its central location for monitoring many sensory input channels and
its collaterals bringing inputs from the cortex and elsewhere.
The reinforcement mechanisms of the CNS help assure that animals learn to repeat activities that bring optimal arousal and
to avoid activities that produce the aversive consequences of
over- and underarousal (Ellis, 1973). The thesis of the present paper is that exploration, play and creativity are among
those types of *operant behavior* that primates learn in order
to generate optimal arousal levels.

There are various properties of *SS* that can induce central
arousal. Each can be conceptualized on a continuum from low,
to medium, to high arousal-induction potential. Any given
stimulus can have several of these properties; and the total
effect of all the properties can be conceived of as the total
"impact" of the stimulus on central arousal (Fiske and Maddi,
1961). *Stimulus intensity* can vary from subliminal, to intermediate, to over-intense; and holding other stimulus properties constant, the extreme levels of intensity are negative
reinforcers while the intermediate levels are positive reinforcers. *Novelty* and *stimulus variety* are also positive stimulus qualities at intermediate levels, whereas extreme novelty
and extreme monotony are aversive. *Stimulus unpredictability*
and *stimulus complexity* are also positive at intermediate
levels. *Cue stimuli* (UCS's, CS's, S^D's) that elicit or evoke
behavior of intermediate intensity and complexity become positive reinforcers, because of the association between the cues
and the reinforcing levels of stimulation.

Stimulus-seeking behavior is shaped by its consequences.
If an exploring *P* discovers a novel stimulus with intermediate
total impact, it will be reinforced for lingering with the
stimulus to explore or play with it. On the other hand, if
the stimulus is so novel or intense that it overarouses the *P*,
the *P* will be punished for continued contact and, thus, negatively reinforced for escaping and avoiding the stimulus.
Finally, non-novel, highly predictable stimuli should not

reinforce much exploration or play unless there are no alternative sources of higher *SS* available in the environment.

Novelty, unpredictability, complexity, and, to some degree, intensity are transactional concepts (Welker, 1971). They must be defined in terms of the stimulus and the *P*'s prior interactions with it. A stimulus may be novel on the first several encounters, but due to *familiarization* and/or *habituation*, it loses its novelty with repeated contacts. Unpredictable events become more predictable with experience. Since complexity correlates with the amount of novelty and unpredictability within a larger stimulus, it, too, provides less arousal after repeated contact and familiarization. In addition, *Ps* can habituate to higher levels of stimulus intensity after repeated contact and, hence, find them less arousing after time. Thus, it is common to find a *P* explore or play with a stimulus during its first contacts (while the stimulus is still novel, optimally arousal-inducing and, hence, a positive reinforcer), then lose interest and turn to more novel alternatives after the first stimulus loses its *SS* reinforcement properties.

Recovery effects operate in the opposite direction from familiarization and habituation. After a *P* has explored or played and become somewhat familiar with a stimulus, a period of separation from the stimulus allows the stimulus to regain a portion of its original novelty and impact for the animal. With renewed impact, a stimulus can again provide positive *SS* reinforcers and, hence, maintain a period of secondary *e&p* until familiarization again robs the stimulus of its novelty and its positive reinforcing properties. Thus, after a period of separation from a stimulus, there is a *rebound* of *e&p*, then a return to baseline levels of interaction (Baldwin and Baldwin, 1976).

Although recovery effects counteract the processes of familiarization and habituation to some degree, the net effect of several presentations and separations from a stimulus is increased total familiarization and loss of *SS* impact. The inevitable loss of impact that accompanies experience normally serves to motivate further exploration and broadening of horizons (up to a limit). As the stimuli in the *P*'s immediate environment become familiar and boring, the *P* is negatively reinforced for escaping by venturing out into larger spheres of its environment; then, if new sources of *SS* are found out there, it is positively reinforced for its new habits of venturing forth. Thus, the process of familiarization plays a key role in motivating the *P* to explore and expand its experiential world. Without familiarization, *Ps* would always find their neonatal environment and activities novel, hence would miss a major motivation for giving them up and would be retarded at leaving behind their infantile ways. Neonatal

activities would remain reinforcing all through life and serve
to hold the organism in an infant stage of psychosocial development or at least serve as competing responses that retarded
the broadening of horizons.

II. THE ONTOGENY OF EXPLORATION AND PLAY

The newborn P has a repertoire of numerous reflex-like
activities. The infant crawls, roots, sucks, clings, rights
itself, startles in response to specific stimuli, and so
forth. These reflex-like behaviors start the infant in life
and provide the raw material from which later behaviors will
be shaped by differential reinforcement, in conjunction with
maturational processes. For the neonate, the world is brand
new; hence, any of the reflex-like responses expose it to new
experiences and begin the process of familiarization. Since
most infant Ps cling to their caretakers, the caretaker's body
is among the first stimuli to become familiar and to lose its
arousal-inducing properties. In addition, the warm, soft
qualities of the caretaker's body are low intensity stimuli
with low arousal-inducing abilities and positively reinforcing
thermoregulation properties. Mason (1965, 1968, 1971) has
clearly demonstrated the functional importance of the low
arousal properties of caretakers. Whenever an infant ventures
away from the caretaker and becomes overaroused by extremely
novel stimuli, it can return to the caretaker for arousal-
reduction and, hence, terminate aversive overarousal. Having
a source of arousal-reducing stimuli is, in fact, crucial for
normal exploration of the environment (Bronson, 1968; King,
1966; Rosenblum, 1971).
Whenever any of the infant's reflex-like behaviors expose
the infant to reinforcers, those behaviors become strengthened
by the reinforcers they produce. Thus, reflex-like crawling
becomes reinforced into operant crawling sequences when it
leads to access to the nipple and milk, to SS or to other
reinforcers. The first case could be called "crawling in the
service of nursing reinforcers"; the second, "crawling in the
service of SS reinforcers"; and so forth. Crawling for no
other purpose than the consumption of optimally arousing SS
corresponds to pure exploration. Exploration can easily expose the infant to overarousing levels of sensory input, such
as contact with a rowdy older individual (A in Figure 3),
which, in turn, is a negative reinforcer. This type of exploration becomes inhibited by the aversive consequences; and
infants learn to some degree to discriminate and not to explore certain subsets of the environment. Once overaroused,
the infant will find sources of arousal-reducing stimuli --

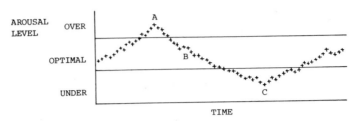

FIGURE 3. During arousal-inducing activities, such as
e&p, a P may become overaroused (A), at which time it usually
seeks arousal-reducing activities, such as clinging to its
mother. After calming down into the optimal arousal zone (B)
or the underarousal zone (C), the P will again find arousal-
inducing activities such as e&p positively reinforcing.

such as the caretaker -- positively reinforcing since they
help counteract the overarousal effects. Because there is
inertia in the arousal system, the infant's internal arousal
levels do not drop down to low levels the instant it contacts
its arousal-reducing caretaker. It may take several minutes
of quiet contact before the infant becomes calmed into its
optimal zone (B in Figure 3) or into the suboptimal zone (C
in Figure 3). In either of these two calm states, the infant
will again find arousal-inducing sensory input positively reinforcing; hence, stimulus-seeking behaviors (e&p) are reinforced at this time. In addition, if the infant's arousal has
fallen into the suboptimal level, the aversiveness of boredom
adds an extra negative contingency to motivate the escape from
boredom into some more stimulating behavior. The frequent al-
ternation between arousal-inducing e&p and arousal-reducing
contact with the caretaker explains why bouts of e&p are often
interrupted with breaks in young Ps.

As the P grows older, several changes occur that make it
less necessary to interject arousal-reduction pauses in play.
1) Although younger Ps react to perturbations in their
arousal level, older Ps often learn to proact in order to
avoid aversive arousal levels (Figure 4). Rather than exploring extremely overarousing stimuli or playing beyond the
point of overarousal then reacting when it is too late, the
P learns to avoid overarousing activities before becoming
completely overstimulated (i.e., to proact) and, hence, to
keep from overshooting the optimal zone. Proacting skills are
also gained as the animal learns to avoid aversive underarousal. Proacting is reinforced because it keeps the P in
its optimal zone a greater percent of the time; but not all
individuals learn to proact equally well. 2) As the young P

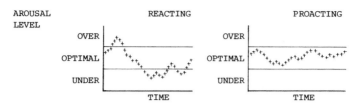

FIGURE 4. *Young Ps tend to react to the conditions of over- and underarousal after experiencing the aversive consequences (left). As Ps learn skills for proacting, they can stay within the optimal arousal zone and experience positive reinforcers a higher percent of the time (right).*

explores an ever broader range of its environment, the stimuli lose their impact on its arousal system via familiarization and habituation. For the neonate, everything is novel, and exploring the caretaker's body is a source of new experience. As that becomes familiar, the infant begins reaching from the caretaker's body for nearby leaves. As the leaves and twigs lose their novelty, the infant is well on the way to accommodating to the higher SS level it will find upon leaving the caretaker to explore through the trees and lianas. After accommodating to higher levels of SS in the nonsocial environment, the infant may explore the social environment and progressively more active types of social play: wrestling, chasing, then rougher play fighting. The processes of familiarization and habituation allow the infant to find higher levels of SS to be optimally arousing as it gains experience (Figure 5). Once the juvenile has accommodated to the higher SS levels of rowdy play, it is less likely to be overaroused by play than the several week old infant would be. Hence, the juvenile has less need of its caretaker for arousal-reduction. The juvenile can stay in play bouts for long periods without breaks, enjoying the high levels of SS provided by jumping, tumbling, wrestling, chasing, and so forth. Pauses in play can occur if an older, stronger player suddenly appears and overstimulates the juvenile; but many of the pauses in play will relate to other aversive contingencies -- such as overheating, fatigue, a painful fall or bump -- which reinforce the temporary termination of the interaction.

The processes of familiarization and habituation are the primary factors that stimulate the escalation in optimal SS level and e&p seen in Figure 5. Numerous factors determine how high the optimal SS zone will rise. In large groups and/ or complex environments, there are high levels of SS to reinforce a longer escalation period and a higher optimal SS

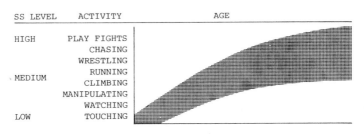

SS LEVEL   ACTIVITY   AGE

HIGH   PLAY FIGHTS
              CHASING
              WRESTLING
 MEDIUM   RUNNING
              CLIMBING
            MANIPULATING
              WATCHING
LOW   TOUCHING

*FIGURE 5. After repeated experience with arousal-inducing
stimuli, familiarization and habituation operate to raise a
P's optimal sensory input level. SS above and below this optimal zone are aversive; SS within this zone is a positive reinforcer. The figure shows the general trend in the awake P's
optimal input zone, averaging out the hourly and daily fluctuations (due to diurnal cycles, fatigue, differences in current
behavior patterns, etc.) in order to focus on broader developmental patterns.*

preference than in sensory restricted environments. However,
if the environment is extremely arousing or the P does not have
an effective source of arousal-reduction, e&p will be inhibited by the aversive consequences of unescapable arousal and the
escalation process retarded (Rosenblum, 1971). Since it takes
behavioral or discriminative (cognitive) skill to extract many
kinds of SS from the environment, Ps who have learned a larger
repertoire of skills have a better chance of succeeding (and
getting reinforcers) in e&p than less skilled peers, hence, of
escalating faster. To the degree that Ps are reinforced for
spending time in competing responses (foraging, predator avoidance, etc.), they will have less time for e&p, although the SS
involved with more arousing competing responses can cause
a rise in optimal SS level, also. Social reinforcers can
strengthen or inhibit e&p along with the skills or the competing responses that indirectly affect e&p. Social models can also affect the e&p, skills and competing responses of onlooking
Ps via observational learning and vicarious reinforcement. For
a more complete discussion of the multiple contingencies that
affect the development of e&p and the life cycle changes in
optimal SS level, see Baldwin and Baldwin (1977).

III.   DEFINING SIX PHASES OF EXPLORATION AND PLAY

     Arriving at an acceptable definition of play has been
problematic (Beach, 1945; Bekoff, 1972; Fagen, 1974; Groos,
1898; Loizos, 1967). Play is one of the most variable of

behaviors in most species' repertoires, and the large number of complex behavioral topographies has made it difficult to base definitions on descriptions of topography alone. An analysis of the *controlling factors* that determine the topography of play might be more profitable than an analysis of topographies alone for defining, predicting and controlling play. Not that topographies are unimportant: a study of controlling factors must be able to explain the topographies displayed. However, a focus on controlling factors allows one to cut through the complexity of highly varied behaviors to the antecedent causes (which *may* be simpler, although not necessarily). This strategy has been extremely successful in the experimental analysis of the proximal causes of behavior (Skinner, 1969), and may be equally useful in studying the ultimate evolutionary causes (Skinner, 1969:183).

Up to this point, we have defined pure *e&p* as a subset of stimulus-seeking behaviors reinforced by optimal levels induced by intermediate *SS*. A more detailed specification of the reinforcement contingencies will allow us to define subsets of both exploration and play. The relationship between *SS* reinforcers and *e&p* can be seen in the behavioral changes that occur from a *P*'s initial exploration of a highly novel target stimulus to the termination of *e&p* with that same stimulus. The typical pattern can be divided into six general phases, based on six different reinforcement contingencies, although much variance can occur within each phase and finer discriminations are possible. (Not all the following phases appear if the stimulus is not sufficiently novel, if the *P* is too young, too old, or too inhibited to play, or if competing responses interfere with the typical pattern.) All six phases can appear in both social and nonsocial forms of *e&p*.

A. Early Exploration (Phase 1)

When a *P* (who has already learned a repertoire of *e&p* behaviors) first encounters a highly novel stimulus, proximity to the stimulus can easily overarouse the *P* which would reinforce moving away. At larger distances, the novel stimulus displaces a smaller portion of the total perceptual field, and -- if the rest of the field is not overarousing -- the novel stimulus is less frightening at this "safe" distance. (If the whole field is extremely novel, the *P* may flee the field or break into distress responses -- such as huddling, rocking or vocalizing. Extreme overarousal and fear can produce such strong avoidance responses that the *P* never familiarizes or habituates to the stimulus and never proceeds to the next phase.) As the *P*'s arousal level drops due to familiarization and habituation, the *P* will find looking at the stimulus a

source of novelty, and may approach for closer examination.
In phase 1, there are often approach-withdrawal responses as
the P vacillates between overarousal and lower arousal states.
As the P further familiarizes and habituates to the stimulus,
the P may scamper up to briefly touch the stimulus, then jump
away after making contact. With repeated contacts, the stimulus continues to lose some of its extreme novelty, the P
ceases to be so easily overaroused by it, and contacts become
more lengthy, although still tentative and interrupted by
bouts of withdrawal.

Early contact with peers at the period before Ps discover
social play has this same approach-avoidance quality with tentative, interrupted contacts.

B. Intermediate Exploration (Phase 2)

As the P becomes familiar with the stimulus, the stimulus
ceases to overarouse and begins to produce optimal arousal.
The P is reinforced for directing its full attention to the
stimulus and for examining it from all different directions if
it presents new stimulation from different angles. As a consequence of the P's investigations and inspection, the P
learns the properties of the stimulus and furthers the processes of familiarization and habituation. As the novelty and
arousal-induction potential of the stimulus decline, there are
diminishing reinforcers for exploration and gradually the P
will show a declining number of exploratory responses, while
phasing into either of the next two phases (Figure 6).

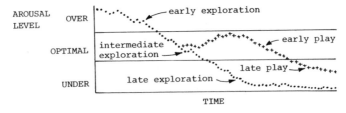

FIGURE 6. The six phases of e&p are determined by the
arousal-induction properties of the stimulus and by the P's
response to the stimulus: during exploration, the P is focusing on what the stimulus is or does; during play, the P is
doing things with or to the stimulus and, thus, creating
heightened stimulus impact.

C. *Late Exploration (Phase 3)*

Once the target stimulus can no longer induce optimal arousal levels, the P is most likely to avoid the stimulus and seek SS elsewhere. However, if a) there are no alternative stimuli with more impact than the target stimulus, and b) the P can no longer succeed in increasing the impact of the target stimulus with its present repertoire of play or creative behavior, then the P may explore the target stimulus as the best source of SS among even worse alternatives. Caged Ps are often forced into this type of exploration. Late exploration tends to be low key, halfhearted and easily interrupted by distractions. Also, the P may not look at the target stimulus as intently as it did in early and intermediate exploration.

D. *Early Play (Phase 4)*

As the target stimulus loses its novelty, the P can create increased SS by *doing* things with or to the stimulus: rolling it, dropping it, hitting it, chasing it, etc. This behavior elicits more SS from the stimulus than is available during much of exploration, in which the P merely attended to the properties of the stimulus. Naturally, a playful swat may cause the stimulus to generate overarousing levels of SS -- if the stimulus is an object that crashes to the ground or a larger animal that startles, jumps up and chases the P -- and this may inhibit the development of playful responses to the particular stimulus.

Primates may spend more time in phase 4 with other animals, especially peers, than with inanimate objects.   The P's repertoire of playful behavior may be able to elicit several dozen new sights and sounds from an inanimate object; but because animals -- especially peers -- often return play, animals may generate much more SS than inanimate objects when played with. Animate objects are usually more complex, more unpredictable and, hence, a less easily exhausted source of SS.   Still, access to the SS they produce is usually contingent upon the P's generating playful behavior; hence, the general rule still obtains: in phase 4, the P must combine its own behavior with the properties of the stimulus to obtain optimal arousal.

Thus, early play tends to be a more active process than exploration, which fits with Welker's (1961) definition.   And compatible with Hutt's (1966) observations, play involves the P's using or doing something with the stimulus; whereas exploration is the prior period in which the properties of the stimulus are learned.   The P at play learns behaviors or games that combine the stimulus with other stimuli, behaviors and play patterns to create more SS than the stimulus alone gives off.

E.   Late Play (Phase 5)

After a variety of play behaviors have been explored and
tried on the target stimulus, the novelty is eventually exhausted. If there are other novel stimuli in the environment,
the *P* may leave the first stimulus and never reach phase 5.
However, if there are no alternatives more novel than the target stimulus, the *P* may continue halfhearted, low key play with
the stimulus merely because it offers at least some *SS* which is
better than nothing. Older juveniles and adults often play in
this manner. They have exhausted most of the readily available
SS in the environment, yet are habituated to relatively high
levels of *SS*. They are bored doing nothing but excited by progressively fewer alternatives in their environment. As they
wait for something new to happen, they put some minimal effort
in the old familiar play patterns. The *SS* of having a young
juvenile jumping around at arm's length can be maintained with
a few swats or incipient chases by the older *P*. The game is
not novel, but the *SS* it generates is better than nothing.

F.   End of e&p *(Phase 6)*

As the target stimulus loses all novelty and arousal-induction capacity, both *e&p* will cease, except for possible re-
bounds of *e&p* after recovery periods.

IV.   CREATIVITY AND PSYCHOSOCIAL GROWTH

Creativity parallels the classification of *e&p* in Figure 1.
Creativity in the service of food getting, escape from pain or
other reinforcers consists of generating a new behavior pattern
for obtaining the relevant reinforcers. "Pure" creativity (in
the service of *SS*) consists of generating new behavior patterns
for *SS* reinforcers. Since novel behavior results from all
forms of creativity, optimal arousal reinforcers may be involved in all types of creativity if the behaviors are sufficiently novel. We will focus on types of creativity related to
SS reinforcers, although admixtures of other reinforcers are
possible.
There are different kinds of creativity, resulting from
different contingencies. First, there are the contingencies
that shape the basic operant behavior repertoire of the infant
as it learns *e&p* behaviors. For the neonate or infant who is
still learning a repertoire of *e&p* activities, many of its
early *e&p* behaviors are new and, hence, creative. This type
of creativity can occur during any of the first four or five

phases of e&p above. In this sense, the young infant is in the most creative period of life because it has so many new behaviors to learn -- even the basics of locomotion, approach-withdrawal, manipulation, and so forth, are new.

There is a second set of contingencies that augments the first and operates even after the infant has learned a basic repertoire of e&p. In phase 4 (and, to a lesser degree, in phases 2, 3 and 5), the P must generate behavior in order to get continued SS reinforcers from the stimulus. With a new stimulus, the P can obtain novel stimulation contingent upon using several of its basic play skills. This may not be the most innovative type of creativity, but it is new in the sense that it recombines old behaviors with a new stimulus for new SS results. Third, an even higher level of creativity occurs when the P generates a new response to the stimulus -- going beyond its previous behavioral repertoire due to response generalization, random variation, or so forth. This new behavior with the new stimulus potentially could produce very novel and/or unpredictable results (although not necessarily positively reinforcing).

Fourth, social contingencies can affect creativity. If the P sees a social model respond to a stimulus in a way the P has never done, the P may learn the new response via observational learning and, hence, be creative, although as a "copy cat". Since an observer seldom copies a model's behavior with 100% accuracy, imitative behavior is more creative than the word imitation alone implies. Social reinforcers can also influence creativity by rewarding a P's new responses. A more creative play partner creates SS for itself and its partners, and, thus, would be more preferred by others for play interaction. The increased social contact and play, because it is contingent on the P's creativity, might further reinforce creativity. In humans, the social rewards we give to some creative individuals doubtlessly help reinforce their creativity.

A review of the literature on e&p revealed that at least 30 functions have been attributed to e&p (Baldwin and Baldwin, 1977), and all but two[3] of these directly serve to facilitate psychosocial growth, helping the individual gain experience and skill, realize its potentialities, and become a socialized member of its group. Ps that are deprived of play can develop an adaptive modicum of behavior compatible with survival in difficult times; but e&p enlarge the P's experiences and

---

[3] *The two functions of e&p that do not directly serve to enhance psychosocial development are (#1) physical exercise and (#30) the benefits that accrue to an abandoned infant because the P had engaged in play-mothering and might adopt that infant.*

behavior repertoire beyond the level learned via food, sex,
thermoregulation and other reinforcers (Baldwin and Baldwin,
1973, 1977).

V. DOVETAILING WITH BIOLOGY

In spite of the emphasis on learning, the present theory
does not neglect biological determinants of e&p (and conse-
quent creativity and psychosocial growth). Tinbergen (1963)
and others have emphasized that behavioral theories need to
integrate evolutionary history, adaptive significance, development and immediate causation.

The adaptive significance of e&p is captured in part by
the 30 functions or adaptive consequences that have been attributed to e&p. The adaptive consequences of e&p doubtlessly
gave a selective advantage to Ps that possessed the behavioral
mechanisms which mediate the development and performance of
e&p. Doubtlessly, some of the 30 adaptive consequences carried heavier causal weightings than others in the selection of
e&p mechanisms, but little data are available on the evolutionary history of e&p with which to estimate those weightings.

The immediate causes of e&p are located in the physiological mechanisms, their development and their interaction with
learning. The physiological mechanisms underlying e&p determined by natural selection are numerous. The reticular formation and diffuse thalamic nuclei have been implicated as the
CNS homeostat that measures arousal levels (Jasper, 1958;
Lindsley, 1951, 1961; Samuels, 1959). The close association
of the ascending reticular system with the hypothalamus and
hippocampus would allow the homeostat to regulate the reinforcement centers of those areas (Ito, 1966; Stumpf, 1965).
Cortical elaboration and species' intelligence affect e&p by
establishing the level of behavioral and discriminative capacity of a species, hence the amount of SS it can locate in a
given stimulus. More intelligent species can discriminate
finer levels of stimulus complexity and invent more creative
ways of treating a given stimulus (Menzel, 1969).

Nutrition, injury and other developmental factors affect
the ontogeny of e&p. Maturation sets limits on the sensory-
motor capacities of the P, hence affects its success or failure (reinforcers) at e&p. Body size, weight, limb structure
and other species-specific traits all set constraints on the
species' behavioral repertoire and affect the type of e&p that
will be reinforcing in a given species (Baldwin and Baldwin,
1977).

The biological study of these adaptations, structures and
their development can advance our understanding of e&p con-

siderably and further the interconnectedness of biological and learning determinants of behavior.

VI. EXPERIMENTS TO BE DONE

    At present, reinforcement theory is relatively unexplored in the primate research on exploration, play and creativity (ep&c). There are, however, many experiments that could be carried out to test it, modify it and expand on it.
    (1) If ep&c are, in fact, operants, one should be able to increase or decrease the frequency of any of these behaviors by use of contingent positive or negative reinforcers. Electrical stimulation to the reinforcement centers of the brain (ESB) might be a convenient methodology[4] since the reinforcers could be delivered while the Ps were engaged in natural behaviors. However, other types of reinforcers have been used in similar experiments in porpoises and humans (Goetz and Baer, 1973; Pryor, Haag and O'Reilly, 1969).
    If ESB or other unobtrusive methods[5] could be used to reinforce the frequency of ep&c, several additional experiments could be run. (2) Reinforcing a P for overselecting toy A should hasten familiarization with that toy and produce a heightened relative preference for toys B, C and D compared with A when the artificial reinforcers are terminated. (3) Chronic augmentation of ep&c should hasten the rise of the optimal SS level (Figure 5) such that the P would appear to be precocious in open field tests or measures of approach to very novel or arousing test stimuli. (4) Chronic punishment of ep&c should have the opposite effect from experiment 3, even for Ps living in a sensory rich environment. (5) Contingent reinforcement of imitation, especially imitation of ep&c, should increase the frequency of the imitator's adopting the ep&c patterns of other Ps, even after the termination of contingent reinforcement. (6) Reinforcement contingent upon taking the role of play initiator or play terminator should affect the frequency of adopting these roles. (7) Some highly innate behaviors -- such as vocalizations -- are resistant to control by reinforcers. One could experiment to determine the degree to which

[4] Beninger, Bellisle and Milner (1977) have shown that ESB reinforcement is more effective than had previously been thought, especially if priming can be avoided.
[5] Clicks or tones associated with food would become secondary reinforcers which might be used if they did not interfere with ep&c.

metacommunicative signals come under the control of reinforcers (as suggested by Baldwin and Baldwin, 1977:379-380).

(8) Contingent positive or negative reinforcers for competing responses should inhibit or disinhibit *ep&c*. Again, ESB
might be used, but other means can be devised to alter the
frequency of competing responses (Baldwin and Baldwin, 1976;
Oakley and Reynolds, 1976). (9) Since guidance and prompts
increase the rate of behavioral change of reinforced operants
(e.g., Fouts, 1972), experiments could test whether *ep&c* are
affected by these procedures. For example, one could modify
the research on primate art (Brewster and Siegel, 1976; Morris, 1962; Rensch, 1973) by adding guidance and prompts coupled with contingent reinforcers, designed to produce specific,
predicted art forms. (10) Central arousal level could be manipulated with drugs, with ESB in the reticular formation and
related areas, or by varying the background sensory stimulation level in the environment. The organism would be predicted to be most reinforced for producing its own *SS* -- via *ep&c*
-- when the experimentally induced arousal was low; and least
reinforced when the experimentally induced level was high.
(One would have to control for habituation and familiarization
effects). (11) Chronic play deprivation -- using ESB, distraction (Oakley and Reynolds, 1976), starvation (Loy, 1970),
or slow feeding apparatuses (Baldwin and Baldwin, 1976) --
would test whether animals deprived of social play (but not
deprived of social contact in an intact group) would develop
adaptive or maladaptive behaviors (Baldwin and Baldwin, 1977).
(12) Deactivation of the pain systems in the CNS should remove
the natural contingency that conditions *P*s not to bite hard in
social play activities. One of the results might be an increase in these pain associated forms of *ep&c*. (13) A rebound
of extra play after a period of play deprivation could be explained by hydraulic or reinforcement theories (Baldwin and
Baldwin, 1976). Reinforcement theory would predict some
events not predicted by hydraulic theories. After the termination of play deprivation, a rebound should occur if there is
low *SS* in the environment but should be progressively weaker
when environmental *SS* is at higher levels. (14) Since skill
should give a *P* an advantage in locating more *SS* in an environment, one could condition an animal to master several
useful (but not normally learned) skills related to *ep&c*, then
measure the degree to which the animal exhibited more *ep&c*.
(15) Since creativity depends to some degree on being able to
recombine old stimuli into new patterns, one might contingently reinforce recombinational behavior, then test to see if
*P*s were more creative in a later test period. (16) Prior positive or negative experience with a stimulus should influence
a *P*'s readiness to approach and explore similar stimuli. One
could devise a box to be explored, such that manipulation of

different parts of the box provided novel stimulation. One could precondition Ps with positive or aversive conditioning to stimuli similar to those on the box, predicting that animals with positive preconditioning would be less reluctant than controls to explore the box, and *vice versa* for animals with aversive preconditioning. (17) If it is true that elevated plasma testosterone levels give males an advantage over females in exploring stimuli that require strength for successful interaction (Baldwin and Baldwin, 1977), experimental injections of testosterone to a test P (male or female) should make available those realms of *ep&c* requiring brute strength. On the other hand, extra strength may increase the likelihood that strong players hurt weaker ones, thus creating punishment contingent upon play and inhibiting weaker animals from playing with the stronger, testosterone-injected ones (Joslyn, 1973). (18) One could study the degree to which female play would parallel male play if testosterone titer were held at male levels in a female matched with a similar weight male. (19) Ps with little access to a source of arousal-reduction (mother, surrogate, etc.) would be expected to exhibit less *ep&c* and to bar press more for access to sources of low level *SS* than controls. (20) The present theory predicts that hypothesized motivations such as a "need for mastery", "competence motivation" and "control over the environment" (Hendrick, 1942; Skinner, 1968; White, 1959) are inferred from normal *ep&c* shaped by *SS* reinforcers. The mastery or competence motive is usually inferred when animals actively manipulate the environment rather than remain passive in it. Being active in generating an external stimulus event creates more proprioceptive *SS* than passive perception of the same external input. If an animal can gain access to a stimulus display via either active (mastery) or passive routes, one would predict a) a preference for the active mode when the P is in low arousal conditions, b) a preference for the passive mode at high optimal arousal levels, and c) avoidance of both modes when the P is above the optimal arousal zone.

VI. HUMANISTIC IMPLICATIONS

    Humans differ from the other primates in numerous ways -- such as language use, greater reliance on observational learning, greater cognitive and discriminative capacity, more complex cultural environments, and so forth. Takine these differences into consideration, one can extend the present primatological theory to include humans.

    There is already a large body of data on *ep&c* in humans that supports the reinforcement and social learning theories proposed above. From the first weeks of life, human infants respond to *SS* as a reinforcer (Berlyne, 1958; Bower, 1966; Leuba and Friedlander, 1968; Watson, 1966). Human *e&p* follows the same patterns as stimulus-seeking behaviors described in primates (Hutt, 1966; Rheingold and Eckerman, 1970; Rheingold, Stanley and Cooley, 1962). Intermediate levels of *SS* are positive reinforcers, whereas high and low levels are negative (Bexton, 1953; Scholtz and Ellis, 1975; Smith, Myers and Murphy, 1962; Switzky, Haywood and Isett, 1974; Vitz, 1966; Zuckerman, Levine and Biase, 1964). Social learning variables (models, social reinforcers, rules and prompts) all affect the rates and topographies of *ep&c* (Buell *et al.*, 1968; Feitelson and Ross, 1973; Goetz and Baer, 1973; Goetz and Salmonson, 1972; Griffing, 1974; Lovinger, 1974).

Several humanistic implications for increasing *ep&c* and psychosocial growth in humans flow directly from reinforcement and social learning theories of stimulus-seeking behaviors.

By providing opportunities for high levels of readily available *SS* in a given environment, there should be more *SS* reinforcers to condition *ep&c*. A high potential stimulation environment would be one in which many activities -- music, sports, art, literature, conversation, etc. -- were contingently available if a person wanted to engage in them, but none were forced. To ensure that people could successfully locate *SS* reinforcers in the available activities and not receive negative reinforcers due to failure, people should be continually (or intermittently) given the opportunity to learn increasing levels of skills for advancing further into their preferred activities (as yesterday's or last month's activities become familiar and less rewarding). To some degree, the learning of these stimulus-seeking skills will be motivated by the *SS* reinforcers they produce; however, social reinforcers, vicarious reinforcers from enthusiastic models and self-reinforcement (Bandura, 1977) can serve as additional means for conditioning those skills. Also, care should be taken to minimize social contingencies that punish and inhibit *ep&c*. Truly creative behavior (when it is not antisocial), and the skills that are generally precursors of it, should be widely acknowledged as valuable in order to increase appreciation and natural social reinforcement.

A multiple approach to skill learning would be the best way to establish well learned skills and to avoid a rigid molding of a stereotyped set of skills. If everyone learned painting from imitating one model or from one rigid set of rules, there would be more conformity -- hence, less creativity, variance and novelty -- than if multiple training methods were used. Multiple models going through different paces would allow each observer to learn a superabundance of skills, from which each might easily combine a different, individually unique subset (Bandura, Ross and Ross, 1963; Harris and Evans, 1973). Models who show enthusiasm or other signs of positive affect or status bring *vicarious reinforcers* to the observers (Bandura and Walters, 1963); thus, the positive cues the models give off while engaging in *ep&c* can increase the total reinforcement for the observers. Biographies, autobiographies, Maslow's (1950) study of self-actualizers and similar sources often contain symbolic models and rules that can enhance *ep&c*.

The use of multiple models is not the only way to condition variable and creative skills. Providing multiple rules from texts, instruction, and critical discussion can shape divergent thinking. Self-observation can also be a good source of rules when a person is able to identify conditions or behaviors that increase or decrease *ep&c*. Meichenbaum (1973), Meichenbaum and Cameron (1974), Schulman (1973), Skinner (1968), Stein (1975) and others provide examples of the rules and contingencies that one could use to enhance spontaneity, playfulness, creativity, and creative problem-solving. Props which serve as prompts also affect *ep&c* (Feitelson and Ross, 1973). Since each individual will be exposed to a different combination of models, rules, prompts and reinforcement patterns, there would be no fear that people would become carbon copies of each other.

A program which practiced good husbandry of *SS* reinforcers would probably be able to carry people further in the acquisition and maintenance of *ep&c* than a program without concern for husbandry. Novelty dies after overexposure: good husbandry prevents overexposure. The goal of *ep&c* is to attain stimulating, fresh experience, not to exhaust the readily available *SS*. A meteoric rise in *ep&c* and in optimal stimulation level might more quickly exhaust the readily available *SS* in an environment than a paced rise in which (1) recovery effects were periodically capitalized on as one returns to rediscover previously stimulating activities, and (2) the available *SS* and other reinforcers were allocated sparingly for reinforcing stimulus-seeking skills (rather than merely consumed noncontingently for pleasure rather than skill learning).

There is no question that people will learn *ep&c* for other reinforcers besides *SS*. Money, attention, praise, fame, and status have all played some role in conditioning *ep&c* throughout history. However, *SS* has certain advantages over contrived social reinforcers in conditioning *ep&c*.[6] A problem with some behavior modification programs is the lack of generalization: the subject performs well in the training environment (where tokens and praise are used) but his behavior does not general-

[6]*Uncontrived social reinforcers -- such as spontaneous applause, praise, admiration from an appreciative audience -- are different and would avert some of the problems mentioned here.*

ize to the natural environment (where less contrived reinforcers are often the only ones operating). When conditioning
ep&c, it would be desirable to give the trainee sufficient
skills that he could generate enough *SS* reinforcers (the *natural reinforcer for ep&c*) that *ep&c* would not extinguish when
the training contingencies were faded out. Lepper, Greene and
Nisbett (1973) showed that excess reliance on extrinsic rewards
for play had a counter productive effect on play after the termination of extrinsic reinforcement. Naturally, one needs to
know how unsupportive of *ep&c* a person's natural environment is
going to be in order to know how severe the problem of generalization will be. If the trainee will be enmeshed in a social
matrix where people value and reinforce creative endeavors,
playfulness and inquisitiveness, there would be less need to
fear the lack of generalization than if the person lived in an
environment where *ep&c* were punished or inhibited by strongly
reinforced competing responses (such as in some religious sects
which denounce *ep&c* in favor of religious activities).

In the nonhuman primates, it is common to see a decline of
ep&c in adulthood, years before physical infirmity, senescence
or senility set in. The causes are numerous (Baldwin and Baldwin, 1977). Many humans show a similar decline in *ep&c* in the
adult years; but the fact that numerous people stay inquisi-
tive, playful and creative well into old age demonstrates that
a decline in *ep&c* is not an inevitable species characteristic.
The present theory suggests that *ep&c* can remain strong all
through life for people in high *SS* environments, with good
stimulus-seeking skills, good husbandry of *SS* reinforcers,
good health and other related factors. The fact that humans
have created multifaceted, man-made cultures which enrich our
SS environment enormously means that each individual can explore worlds of *SS* via travel, books, conversation, games, music, mass media,[7] crafts, sports, scientific studies, and so
forth, without exhausting all the possible sources of *SS*. For

---

[7]*It should be noted that the mass media provide cheap (low
effort) SS that rewards passive forms of stimulus-seeking behavior, and, hence, does not shape many skills. The person
who explores music or sports by actively playing them reaps SS
reinforcers contingent upon increasing skills, but not the person who explores music or sports via mass media or mechanical
reproduction. As our populace spends an increasing number of
hours in front of T.V., going to high SS movies, or consuming
mass entertainment, their skills for creating their own SS may
go underdeveloped. As we let the machines and entertainment
industry do the work of generating SS for us, we may become
dependent on them and less creative ourselves. It might be
necessary to avoid passive stimulus-seeking if one wants to
come under the contingencies to learn the active forms.*

those who have sufficient skills and health, and are free of
punitive contingencies or competing responses, a lifetime of
ep&c is possible. An increased social concern on fulfilling
the older years could help reinforce ep&c and other good behaviors far into old age. Symbolic models are already there in
biographies, autobiographies, and books on aging (e.g., Comfort, 1976). Since we all will grow old, it might behoove us
to proact.

ACKNOWLEDGMENTS

We gratefully acknowledge Howard Goldstein for his help in
preparing this manuscript.

REFERENCES

Baldwin, J. D., and Baldwin, J. I. The role of play in social
    organization: comparative observations on squirrel monkeys
    (Saimiri). Primates 14, 369-381 (1973).
_____ Effects of food ecology on social play: a laboratory
    simulation. Z. Tierpsychol. 40, 1-14 (1976).
_____ The role of learning phenomena in the ontogeny of ex-
    ploration and play, in "Primate Bio-Social Development:
    Biological, Social, and Ecological Determinants" (S. Chev-
    alier-Skolnikoff, and F. E. Poirier, eds.), pp. 343-406.
    Garland Press, New York (1977).
Bandura, A. "Social Learning Theory". Prentice-Hall, Engle-
    wood Cliffs, New Jersey (1977).
Bandura, A., Ross, D., and Ross, S. A. A comparative test of
    the status envy, social power, and secondary reinforcement
    theories of identificatory learning. J. abnorm. soc. Psy-
    chol. 67, 527-534 (1963).
Bandura, A., and Walters, R. H. "Social Learning and Personal-
    ity Development". Holt, Rinehart and Winston, New York
    (1963).
Beach, F. A. Current concepts of play in animals. Amer.
    Natur. 79, 523-541 (1945).
Bekoff, M. The development of social interaction, play and
    metacommunication in mammals: an ethological perspective.
    Quart. Rev. Biol. 47, 412-434 (1972).
Beninger, R. J., Bellisle, F., and Milner, P. M. Schedule con-
    trol of behavior reinforced by electrical stimulation of
    the brain. Science 196, 547-549 (1977).
Berlyne, D. E. The influence of albedo and complexity of stim-
    uli on visual fixation in the human infant. Brit. J. Psy-
    chol. 49, 315-318 (1958).

————— "Conflict, Arousal, and Curiosity". McGraw-Hill, New York (1960).

Bexton, W. "Some Effects of Perceptual Isolation on Human Subjects". Ph.D. Dissertation, McGill University, Montreal (1953).

Bower, T. G. R. The visual world of infants. *Sci. Amer. 215,* 80-92 (1966).

Brewster, J. M., and Siegel, R. K. Reinforced drawing in *Macaca mulatta. J. hum. Evol. 5,* 345-347 (1976).

Bronson, G. W. The development of fear in man and other animals. *Child Develop. 39,* 409-431 (1968).

Buell, J., Stoddard, P., Harris, F., and Baer, D. Collateral social development accompanying reinforcement of outdoor play in a preschool child. *J. appl. behav. Anal. 1,* 167-173 (1968).

Butler, R. A. The differential effect of visual and auditory incentives on the performance of monkeys. *Amer. J. Psychol. 71,* 591-593 (1958).

————— Investigative behavior, *in* "Behavior of Nonhuman Primates", Vol. 2 (A. M. Schrier, H. F. Harlow, and F. Stollnitz, eds.), pp. 463-493. Academic Press, New York (1965).

Campbell, H. J. Peripheral self-stimulation as a reward in fish, reptile and mammal. *Physiol. Behav. 8,* 637-640 (1972).

Comfort, A. "The Good Age". Crown, New York (1976).

Ellis, M. J. "Why People Play". Prentice-Hall, Englewood Cliffs, New Jersey (1973).

Fagen, R. Selective and evolutionary aspects of animal play. *Amer. Natur. 108,* 850-858 (1974).

Feitelson, D., and Ross, G. The neglected factor -- play. *Hum. Develop. 16*(3), 202-223 (1973).

Fiske, D. W., and Maddi, S. R. "Functions of Varied Experience'. Dorsey Press, Homewood, Illinois (1961).

Fouts, R. S. Use of guidance in teaching sign language to a chimpanzee *(Pan troglodytes). J. comp. physiol. Psychol. 80,* 515-522 (1972).

Goetz, E. M., and Baer, D. M. Social control of form diversity and the emergence of new forms in children's blockbuilding. *J. appl. behav. Anal. 6,* 209-217 (1973).

Goetz, E. M., and Salmonson, M. M. The effect of general and descriptive reinforcement on "creativity" in easel painting, *in* "Behavior Analysis and Education" (G. Semb, ed.), pp. 53-61. University of Kansas Press, Lawrence (1972).

Griffing, P. Sociodramatic play among young black children. *Theory into Practice 13*(4), 257-265 (1974).

Groos, K. "The Play of Animals". D. Appleton & Co., New York (1898).

Harris, M. B., and Evans, R. C. Models and creativity. *Psychol. Rep. 33,* 763-769 (1973).

Hebb, D. O. Drives and the CNS (conceptual nervous system).

Psychol. Rev. 62, 243-254 (1955).

_____ "A Textbook of Psychology", 2nd Ed. W. B. Saunders Co., Philadelphia (1966).

Hendrick, I. Instinct and the ego during infancy. *Psychoanal. Quart. 11,* 33-58 (1942).

Humphrey, N. K. Species and individuals in the perceptual world of monkeys. *Perception 3,* 105-114 (1974).

Hutt, C. Exploration and play in children. *Symp. Zool. Soc. Lond. 18,* 61-81 (1966).

Ito, M. Hippocampal electrical correlates of self-stimulation in the rat. *Electroencephalogr. clin. Neurophysiol. 21,* 261-268 (1966).

Jasper, H. H. Reticular-cortical systems and theories of the integrative action of the brain, *in* "Biological and Biochemical Bases of Behavior" (H. F. Harlow, and C. N. Woolsey, eds.), pp. 37-61. University of Wisconsin Press, Madison (1958).

Joslyn, W. D. Androgen-induced social dominance in infant female rhesus monkeys. *J. child Psychol. Psychiat. 14,* 137-145 (1973).

King, D. L. A review and interpretation of some aspects of the infant-mother relationship in mammals and birds. *Psychol. Bull. 65,* 143-155 (1966).

Lepper, M. R., Greene, D., and Nisbett, R. E. Undermining children's intrinsic interest with extrinsic rewards. *J. Pers. soc. Psychol. 28,* 129-137 (1973).

Leuba, C., and Friedlander, B. Z. Effects of controlled audiovisual reinforcement on infants' manipulative play in the home. *J. exp. Child Psychol. 6,* 87-99 (1968).

Lindsley, D. B. Emotion, *in* "Handbook of Experimental Psychology" (S. S. Stevens, ed.), pp. 473-516. Wiley, New York (1951).

_____ Common factors in sensory deprivation, sensory distortion, and sensory overload, *in* "Sensory Deprivation" (P. Solomon, P. Kubzansky, P. Leiderman, J. Mendenson, and D. Wexler, eds.), pp. 174-194. Harvard University Press, Cambridge, Massachusetts (1961).

Loizos, C. Play behaviour in higher primates: a review, *in* "Primate Ethology" (D. Morris, ed.), pp. 176-218. Weidenfeld & Nicolson, London (1967).

Lovinger, S. Socio-dramatic play and language development in preschool disadvantaged children. *Psychol. in the Schools 11*(3), 313-320 (1974).

Loy, J. Behavioral responses of free-ranging rhesus monkeys to food shortage. *Amer. J. phys. Anthropol. 33,* 263-272 (1970).

Maslow, A. H. Self-actualizing people: a study of psychological health. *Personality Symposia #1,* 11-34 (1950).

Mason, W. A. The social development of monkeys and apes, *in*

"Primate Behavior: Field Studies of Monkeys and Apes" (I.
DeVore, ed.), pp. 514-543. Holt, Rinehart & Winston, New
York (1965).

_____ Early social deprivation in the nonhuman primates: implications for human behavior, in "Environmental Influences" (D. Glass, ed.), pp. 70-101. Russell Sage Foundation,
New York (1968).

_____ Motivational factors in psychosocial development, in
"Nebraska Symposium on Motivation" (W. J. Arnold, and M. M.
Page, eds.), pp. 35-67. University of Nebraska Press,
Lincoln (1971).

Meichenbaum, D. Enhancing creativity by modifying what S's
say to themselves. Unpublished manuscript. University of
Waterloo, Ontario, Canada (1973).

Meichenbaum, D., and Cameron, R. The clinical potential of
modifying what clients say to themselves, in "Self-Control:
Power to the Person" (M. J. Mahoney, and C. E. Thoresen,
eds.), pp. 263-290. Brooks/Cole, Monterey, California
(1974).

Menzel, E. W., Jr. Chimpanzee utilization of space and responsiveness to objects: age differences and comparison with
macaques. Proc. 2nd Int. Congr. Primat. 1, 72-80 (1969).

Morris, D. "The Biology of Art". Knopf, New York (1962).

Oakley, F. B., and Reynolds, P. C. Differing responses to social play deprivation in two species of macaques, in "The
Anthropological Study of Play: Problems and Perspectives"
(D. F. Lancy, and B. A. Tindall, eds.), pp. 179-188. Leisure Press, Cornwall, New York (1976).

Pryor, K., Haag, R., and O'Reilly, J. The creative porpoise:
training for novel behavior. J. exp. Anal. Behav. 12,
653-661 (1969).

Rensch, B. Play and art in apes and monkeys. Symp. IVth Int.
Congr. Primat. 1, 102-123 (1973).

Rheingold, H. L., and Eckerman, C. O. The infant separates
himself from his mother. Science 168, 78-90 (1970).

Rheingold, H. L., Stanley, W. C., and Cooley, J. A. Method for
studying exploratory behavior in infants. Science 136,
1054-1055 (1962).

Rosenblum, L. A. The ontogeny of mother-infant relations in
macaques, in "Ontogeny of Vertebrate Behavior" (H. Moltz,
ed.), pp. 315-367. Academic Press, New York (1971).

Samuels, I. Reticular mechanisms and behavior. Psychol. Bull.
56, 1-25 (1959).

Scholtz, G., and Ellis, M. Repeated exposure to objects and
peers in a play setting. J. exp. Child Psychol. 19, 448-
455 (1975).

Schulman, M. Backstage behaviorism. Psychol. Today 7(1), 51-
88 (1973).

Schultz, D. D. "Sensory Restriction: Effects on Behavior".

Academic Press, New York (1965).

Simpson, M. J. A.   The study of animal play, *in* "Growing Points
in Ethology" (P. Bateson, and R. Hinde, eds.), pp. 385-400.
Cambridge University Press, Cambridge (1976).

Skinner, B. F.   "The Technology of Teaching".   Prentice-Hall,
Englewood Cliffs, New Jersey (1968).

_____   "Contingencies of Reinforcement".   Appleton-Century-
Crofts, New York (1969).

Smith, S., Myers, T., and Murphy, D.   Activity pattern and
restlessness during sustained sensory deprivation.   Paper
presented at American Psychological Association, St. Louis,
Missouri (1962).

Stein, M. I.   "Stimulating Creativity, Volume 2: Group Proceed-
ures".   Academic Press, New York (1975).

Stumpf, C.   The fast component in the electrical activity of
the rabbit's hippocampus.   *Electroencephalogr. clin.
Neurophysiol. 18,* 477-486 (1965).

Switzky, H., Haywood, H., and Isett, R.   Exploration, curiosity, and play in young children: effects of stimulus
complexity.   *Develop. Psychol. 10,* 321-329 (1974).

Tinbergen, N.   On aims and methods of ethology.   *Z. Tierpsychol. 20,* 410-433 (1963).

Vitz, P. C.   Affect as a function of stimulus variation.   *J.
exp. Psychol. 71,* 74-79 (1966).

Watson, J. S.   The development and generalization of "contingency awareness" in early infancy: some hypotheses.
Merrill-Palmer Quart. 12, 123-136 (1966).

Welker, W. I.   An analysis of exploratory and play behavior in
animals, *in* "Functions of Varied Experience" (D. W. Fiske,
and S. R. Maddi, eds.), pp. 175-226.   Dorsey Press, Chicago (1961).

Welker, W. I.   Ontogeny of play and exploratory behaviors: a
definition of problems and a search for new conceptual
solutions, *in* "The Ontogeny of Vertebrate Behavior" (H.
Moltz, ed.), pp. 171-228.   Academic Press, New York (1971).

White, R. W.   Motivation reconsidered: the concept of competence.   *Psychol. Rev. 66,* 297-333 (1959).

Zuckerman, M., Levine, S., and Biase, D.   Stress response in
total and partial perceptual isolation.   *Psychosomat. Med.
26,* 250-260 (1964).

OBJECT-PLAY: TEST OF A CATEGORIZED MODEL BY THE
GENESIS OF OBJECT-PLAY IN *Macaca fuscata*[1]

Douglas K. Candland
Jeffrey A. French

Program in Animal Behavior
Bucknell University
Lewisburg, Pennsylvania

Carl N. Johnson

Department of Child Development and Child Care
University of Pittsburgh
Pittsburgh, Pennsylvania

I. INTRODUCTION

Play is of surprisingly recent interest in the history of
the behavioral sciences, although it has long been of concern
to Western thought, as Smith describes in his introductory
chapter in this volume. A mode of behavior that has been both
condemned and honored, it has mostly been viewed as an activity that, while interfering with production and encouraging
unwanted imagination, is nevertheless necessary to maintain
peak performance or to develop adult skills. Our educational
and religious institutions remain uncertain as to the status
of play, an ambivalence relfected in the questions and strategies we use to study it (Aries, 1962; Symons, this volume;
Veblen, 1934).

[1]*This work is dedicated to Frank DuMond (1930-1977), whose
development of Monkey Jungle and generosity with its facilities provided many investigators with rare opportunities for
research alongside his own insightful understanding and contributions toward the study of primate behavior.*

The last decades of the nineteenth century yielded the
first comprehensive analyses of play. Karl Groos, whose thor-
ough analyses remain current, published his masterful volumes
in 1898 and 1901 (translated into English by the pioneer evo-
lutionist, Elizabeth Baldwin, in 1899 and 1907). Spencer's
two-volume work on psychology that was published in 1873 included a section on the psychology of play. Within these
works are found the ideas that continue to influence our thinking about play, although much that is contained therein has
been innocently rediscovered from time to time. Among the
Grundlagen are these: play is most suitably explained in evolutional and, hence, phyletic terms; play is more common among
higher forms of life, presumably because only the higher forms
have time for it; play represents the discharge of excessive
energy; play is learned through imitation; the degree and kind
of play are related to a species' ability to learn; and play
is preparation for adult life, including the important shift
made ontogenetically from homosexual groups to heterosexual
bonds.[2]

The burst of scholarly concern about play on the part of
the Swiss, English, and Americans at the end of the nineteenth
century was almost certainly caused by their growing understanding of evolutional principles and of the possibilities
these principles provided for explaining complex activities
such as play. It is noteworthy that Spencer, Groos, Mark Baldwin, and Hall intended to divorce their discussions of play
from its religious evaluation. This distinction made possible
a more productive categorization of playful activities than
was possible when play was considered in terms of its moral
qualities. An examination of encyclopedias published during
the last century shows an abrupt change in scholarly attitudes
toward play. Until Groos' work, play was described as a form
of leisure, and leisure was described in terms of its effect
on productivity and religious values. Its nonutilitarian nature was emphasized, implicitly underscoring its danger.
Sometime after Groos' work became known, the ontogenetic and
phylogenetic importance of play became the primary focus of
scholarly work. Mark Baldwin was unquestionably the most capable interpreter of evolution's effect upon behavior, and it
is to his appreciation of Groos' work and his wife's scholar-

---

[2]*The suggestion made by G. Stanley Hall (1906) that because ontogeny is a form of phyletic recapitulation, play of
the human child constitutes behavior of adult hominoid ancestors or less developed races seems to be of little interest
today, although its relationship to the postulate of a collective unconscious (Jung, 1953) is an intriguing idea to some.*

ship that we owe the introduction of Groos to the English-
reading world.

This chapter presents some of the salient difficulties encountered in understanding play, especially primate play, by
focusing (in Section II) on why the comparative psychology of
play, although seemingly a most suitable strategy, has not
been productive. In the third section, we suggest a model of
categories of one possible schematization of play, object-play,
the need for such categorization having hopefully been shown
in Section II.

In the fourth section, we determine how well our provisional categories suit the several elaborations of nonhuman primate
play. In the fifth section, we elaborate these categories by
referring to the extensive literature on play in human infants
and children.

We turn, in Section VI, to a presentation and discussion
of an instructive example of play among *Macaca fuscata*, play
that illustrates the utility of several of our categories. In
this section, we describe how a captive group of *M. fuscata*
came to fashion part of the environment so that this modified
environment then provided for playful activity. We trace
learning of this type of play ontogenetically in order to see
how play is acquired and retained, thereby leading to the conclusions of Section VII.

II. THE COMPARATIVE PROBLEM

If the origins and purposes of play have escaped our understanding, the reason is surely not that play has escaped our
scrutiny. The modern phylogenetic scale shows that play, however loosely defined, is found principally among the Orders
Primates and Carnivora, with sporadic assignments of play to
other isolated species, but not to complete Orders (Smith,
this volume; Welker, 1961).

Inquiry becomes muddled, however, when we can neither define nor accurately describe the behavior under consideration
and when we cannot form categories to suit our tentative descriptions. Primate play should be a fertile field for the
sorts of comparisons promised by comparative psychology. After all, comparative psychology has not found it inconvenient
to compare species of vastly different orders or from quite
distinct evolutionary lines. How odd, then, that when the subject is primate play, a behavioral phenomenon common to an entire Order and, thereby, comparable among genera and closely
related species, the comparisons are not made. This oversight
suggests either that comparative psychology has been blinded

by the richness of the offering or that there is something
about the offering that is suspect.

When we realize that comparative psychology bears strong
epistemological similarities to Aristotelian strategies,[3] we
see the oversight differently. Play, whether primate or non-
primate, has not been ascribed the descriptive categories that
are necessary if comparative psychology, or any other strategy
of comparison, is to be operable. The comparative psychologist can no more work without nomenclature appropriate throughout species than a botanist can work without naming flora.
The naming and categorization exemplified by terms such as
"animal intelligence", "learning ability", or, more recently,
"social structure" are the sources of the power of the system.
If fundamental steps of naming and categorizing have not been
taken, comparisons across behavioral and functional divisions
have no meaning.

We suggest that a comparative psychology of play can provide not only meaningful comparisons, but can also generate
views of play somewhat more productive than those provided by
the present division of structural and functional descriptions.
It is a comforting characteristic of comparative psychology
that while it must first name and categorize in order to compare, the comparisons that it provides after having done so
yield a realignment of the categories first used. The provisional naming and categorization of play thus serve the
function of permitting comparison, with the result that the
comparisons produced lead to a more useful set of definitions
and categories than those conceived initially.

III. A MODEL OF THE CATEGORIES OF PLAY

In his classic work on play in human beings, a work that
followed his perhaps more imaginative work on play in animals,
Groos defended the need to categorize: "Systematization is...
a mere logical ideal. Yet even an imperfect classification
may justify itself in two ways: it may be very comprehensive
and practical, or its aptly chosen grounds of distinction may
serve to open at once to the reader the inmost core of the
subject under discussion" (Groos, 1901:1).

Groos' own categorization is twofold. The first categorization is by sensory system, e.g., whether play stimulates
sight, touch, hearing, and so on. The second is by discipline, e.g., whether one studies the physical properties of
play, its effects on the biological system, or its influence

[3]*Millar (1968) has been generous in ascribing to Aristotle
a clear position on the purpose of play.*

on the feelings produced (the latter being an example, to
Groos, of the psychology of play).

The use of relevant sensory systems as the mode of categorization is helpful in organizing the varied kinds of experiences we label "play", and, indeed, it well suits the idea of
a classification that is "comprehensive and practical". When
such a categorization is completed, nonetheless, it does not
appear to show the "inmost core", for there is neither the
gradation nor the communality among categories necessary for
the model to be instructive as well as inclusive.

We should remember in setting out categories that one does
so as one tries out hypotheses or shoes: each is to be tried
and discarded if it is found wanting. The only justification
for selecting one set of categories in preference to another
is explanatory utility. Our purpose is to suggest categories
that, for the moment, serve to organize what is known about
primate play so the differences and similarities are apparent
and with the expected result that our attention is drawn to
similarities, distinctions, and perhaps causal relationships
that would otherwise have escaped our notice. Our procedure
is to present our categorization and in Sections IV and V to
see how well or poorly our model of categories succeeds. Our
work with *M. fuscata* described in Section VI may then be seen
as an instructive example of the utility of the categories.

We suggest the use of two categories: (1) whether or not
an object is incorporated into the play sequence, and (2) who
or what is involved in the play sequence. The first category
involves two alternatives: an object is incorporated (Object-
play) or is not (Nonobject-play). The second category, which
interacts with the first, provides three alternatives: solitary play, social play,and play with other objects. The 2 x 3
categorization yields six interactions, as shown in Table I.

Two points need to be made regarding the categories of
Table I. First, the category Nonobject-play, coupled with
play-with-other-objects, does not exist, since the two events
contradict one another. Second, there is no obvious place for
behaviors that some have called playful that involve manipulation or exploration of aspects of the environment. An example

TABLE I. *Descriptive Categories and Their Interactions[a]*

	Object-play	Nonobject-play
Solitary	X	X
Social	X	X
Other objects	X	-

[a] X = viable category; - = logically nonviable category

is a monkey's examining and toying with foodstuff. Although such cases involve an animal using an object, it is uncertain that the event is playful, although some definitions would so classify it. Our uncertainty is expressed by our including these cases in a distinct category in Table II, thus making it possible for the reader to judge the proper location of these events.[4]

A third concern is the relationship between our categories and what is known as "tool-use". There are two current definitions of tool-use. The more common, as expressed by Hall (1963:479), is that tool-using involves an animal's manipulating an inanimate object for a purpose (in other words, behaviors described in our model as Object-play/Social and Object-play/Other objects). Alcock's definition of tool-use (1972: 464) is far more stringent, requiring as it does that the object be used to manipulate some additional part of the environment. This definition has the virtue of excluding from tool-using activities the spinning of spiderwebs and the weaving of nests. In our model, the use of Alcock's definition would firmly separate the categories Object-play/Other objects from Object-play/Social. It would appear wise to follow Alcock's lead, both because it is appropriately stringent and because it makes the distinctions between our categories more profound.

In order to test the utility of the categories, we have assembled examples of cases from the literature on primate play. These are shown in Table II. The results of testing the fit of reported studies to the model of categories are informative in several ways.

First, it is evident that the categories serve to separate taxonomic groups, as shown by the progression of categories that have no examples. Although apes and human children show play that fits all of the categories, Old World primates do not appear to show Object-play/Other objects, New World primates show neither that nor Object-play/Social, and prosimians show neither these nor Object-play/Solitary.

Second, the effect of captivity is striking. The only examples of Object-play/Solitary for New World monkeys occur in captive conditions. Moreover, it becomes evident that the

[4]*The problem is fundamental; we tend to regard play as being distinct from manipulation because the latter serves a purpose, while the former does not. But how do we measure or judge purpose? It should not escape our attention that this implicit definition of play is based on the value that playful behavior is somehow unproductive. We have commented before on the damage done when value systems are used to define behavioral events.*

conditions of captivity lead to an increasing number of examples as one reads from prosimians to human children. The
effect of captive conditions on primate play, as deduced from
the model, is to enhance the probability of playful activity
being observed. One possible reason for this is that the ease
of observation may lead to an unduly high number of observations of playful behavior. If it is true that the frequency
of playful behavior is actually greater under captive conditions, a variety of reasons suggest themselves: lack of social contact, boredom, a paucity of objects and, hence, en-
hancement of manipulation of whatever items are available.
Each of these explanations is unsatisfactory, for none is an
explanation so much as a description of some inferred intervening variable.

It would appear that, provisionally, the model succeeds in
separating events into meaningful categories and that it may
serve to show some aspects of what Groos called the "inmost
core" of the phenomenon, as shown by its power in separating
phyletic groups and by the fact that the effects of captivity
are orderly in regard to phyletic status. The next test of
the model is whether observed behavior patterns classified as
playful fit the categories, or whether, once these behavior
patterns are considered, we find legitimate observations without categories and legitimate categories without observations.
The next two sections take up this evaluation, the first in
regard to nonhuman primate play and the second in regard to
human play.

IV. PLAY IN NONHUMAN PRIMATES

One of the notable aspects of primate play behavior is
that much of it is directed toward inanimate objects in the
environment. Certainly, other species play with objects (e.g.,
a puppy chasing a stick, a kitten wrestling with yarn), but it
is only within the Order Primates that one finds complex forms
of object manipulation, the use of an object as an exploratory
tool, and nonstereotyped, novel uses of objects as vehicles
of play.

There are several factors that could account for the diverse uses of objects by primates, including (1) the structural complexity and efficient functioning of the primate hand
(Bishop, 1962), (2) an increased period of infant dependency
on the parent, and (3) the increased capacity for observational learning and cultural transmission of object and tool use.
Regardless of the ultimate causes, the common occurrence of
object play in primates represents a propensity to respond to

TABLE II.

		HUMAN CHILDREN (Hominidae)	APES (Hylobatidea & Pongidae)
OBJECT-PLAY	FOCUS - OTHER OBJECTS	*Constructing a two-block tower, fitting form together. Gessell et al. (1940). Using supports, string and stick to obtain objective. Piaget (1952). Grouping objects, functional relations. Fenson et al. (1976).*	Pan troglodytes; juveniles; fishing for objects with long sticks. Köhler (1925).[a] Pan troglodytes; 2-4 years; play termite fishing. van Lawick-Goodall (1968).
	FOCUS - SOCIAL	*Simple social games of ball, peek-a-boo. Mueller & Vandell (in press). Competition for objects. Aldis (1975). Complex rule-governed games such as marbles and baseball.*	Pan troglodytes; juveniles; tugs-of-war and jumping with poles. Menzel (1972, 1973).[a] Pongo pygmaeus; play with towels and cloths attracts others to join in object-play bouts. Maple (pers. comm.).[a] Pan troglodytes; juveniles; acquired an object-play "tradition" in social triads. Menzel et al. (1972).[a]
	FOCUS - SOLITARY	*Kicking to move a mobile, banging and throwing objects. Piaget (1952). Swings, jump ropes, stilts, etc., as extensions of personal skills. Play with soft, cuddly objects. Freud (1965).*	Pongo pygmaeus; 1-2 years; manipulate substrate, grasses, construct play-nests. MacKinnon (1974).[a] Pan troglodytes; 1-3 years; construct play-nests. Reynolds & Reynolds (1965); van Lawick-Goodall (1968). Gorilla gorilla beringei; 8-24 months; solitary play with a variety of inanimate objects and vegetation. Schaller (1963, 1965). Hylobates syndactylus; 1-3 years; manipulate substrate in playful situations. Chivers (1974). Gorilla gorilla beringei; juveniles & adults; manipulate and play with sticks, dirt clods, and grass. Maple (pers. comm.).[a]
NONOBJECT-PLAY	FOCUS - OTHER OBJECTS		
	FOCUS - SOCIAL	*Rough-and-tumble play. Blurton-Jones (1967). Play-fighting. Aldis (1975). Parent-infant play. Brazelton et al. (1974).*	MANY EXAMPLES. SEE JEWELL & LOIZOS (1966); DOLHINOW & BISHOP (1970); AND BRUNER, JOLLY & SYLVA (1976) FOR REVIEWS OF NONOBJECT-PLAY IN PRIMATES.
	FOCUS - SOLITARY	*Sensory-motor functioning: looking-to-look, sucking-to-suck. Piaget (1952). Fantasy play.*	
	EXPLORATION AND MANIPULATION	MANY EXAMPLES. SEE HUTT (1971); McCALL (1974).	FOR REVIEWS, SEE BUTLER (1965) AND WELKER (1971).
	PRIMATE GROUP	HUMAN CHILDREN (Hominidae)	APES (Hylobatidea & Pongidae)

[a]Reference to captive primates.

Categories of Play and Examples[a]

Old World Primates (Cercopithecoidae)	New World Primates (Cebidae & Callitrichidae)	Prosimians
Macaca fascicularis; juveniles; play with objects is often an invitation for social play. Poirier & Smith (1974). Macaca fuscata; infants & juveniles; prolonged and intense object manipulation is part of social play. Menzel (1966). Cercopithecus aethiops; 1 & 3 years; low frequency of group tugs-of-war, group interest in objects. Fedigan (1972).[a] Papio anubis; 6-12 months; supplement social play by the inclusion of objects. Hall & DeVore (1965). Macaca fuscata; infants; play with sticks, stones and snowballs. Eaton (1972).[a] Macaca fuscata; INFANTS AND JUVENILES: ACQUIRE AND MAINTAIN PLAY WITH STICKS. This report.[a]		
Presbytis entellus; 5-12 months; exploration and play with objects. Jay (1965). Macaca fuscata; adult; constructed snowball. Eaton (1972).[a] Papio anubis; 4-6 months; infants pick up objects systematically. Hall & DeVore (1965). Macaca mulatta; 2-3 months; object-play first occurs near the mother. Meier & Devanney (1974).[a] Theropithecus gelada; juveniles; manipulated rock and balanced them on heads. Bernstein (1975).[a] Macaca nigra; no age given; insert sticks into fencing and then bounce on these. Bernstein, quoted in Beck (1976).[a] Macaca fuscata; JUVENILE: FASHIONED SWING AND USED AS SUCH. This report.[a]	Cebus apella; juveniles & adults; tossing a handkerchief in enclosure. Moynihan (1976).[a] Cebus capucinus; no age given; individual playing with a ball on its tail. Oppenheimer (pers. comm.).[a]	
MANY EXAMPLES. SEE JEWELL & LOIZOS (1966); DOLHINOW & BISHOP (1970); AND BRUNER, JOLLY & SYLVA (1976) FOR REVIEWS OF NONOBJECT-PLAY IN PRIMATES.	Review of New World species. Nonaggressive chases and play-fights. Moynihan (1976).	Review of prosimian species. Nonaggressive rough-and-tumble play, wrestling. Doyle (1974).
	Much locomotor play. See Baldwin (1969), Saimiri sciureus; Stevenson & Poole (1976), Callithrix jacchus; Altmann (1959), Alouatta palliata.	Locomotor play. See Doyle et al. (1969), Galago senegalensis moholi; Jolly (1966), Lemuroid species.
FOR REVIEWS, SEE BUTLER (1965) AND WELKER (1971).	Callithrix jacchus; 4-6 weeks; manipulation accounts for much of daily activity. Rothe (1970).[a] Cebus capucinus; juveniles; examine objects found in trees. Oppenheimer (pers. comm.). Saimiri sciureus; Rosenblum & Cooper (1968).	Tupaia glis, Lemur catta, & L. fulvus; adults; exhibit some manipulative abilities in food-getting situations. Jolly (1964).[a]
OLD WORLD PRIMATES (Cercopithecoidae)	NEW WORLD PRIMATES (Cebidae & Callitrichidae)	PROSIMIANS

objects and a genetic potentiality for using objects in novel and, possibly, beneficial ways, ways apparently requisite for tool-use.

This section characterizes the forms of object-play and manipulation observed throughout the Order Primates, provides a descriptive ontogeny of object-play, and describes the effects of familiarity with objects and of rearing situations on the expression of object-play.

There have been several comprehensive reviews of the literature pertaining to tool-using in nonhuman primates (van Lawick-Goodall, 1970; Warren, 1976), yet to our knowledge, there has been no published review dealing explicitly with *object-play* in primates. Table II is comprised of examples of object-play that provide rich descriptions of it. These examples are categorized according to the model established in Section III and according to convenient primate classifications: Prosimii, Ceboidae, Cercopithecoidea, and Hylobatidea and Pongidae.

One problem in classifying responses to objects is to maintain a distinction between exploration of objects and play with objects. This is not a new problem [see Weisler and McCall (1976) for a discussion of the similarities and differences between the two classes of behavior]. In order to deal with this question, the following criterion was used for classification: responses to an object that were relatively stereotyped among individuals and that showed a high degree of involvement of sensory modalities (e.g., sniffing, tasting, touching) were classified as exploratory responses, while those characterized by a high degree of effector involvement, occurring in apparently low arousal situations and showing a lack of consistency among individuals, were classified as playful responses. Aldis (1975) draws similar conclusions, stating that object-play is a motor activity and exploration is a sensory activity, although both behaviors are likely to exhibit combinations of both sensory and motor components. In our analysis, therefore, whether an activity was sensory dominated or motor dominated was an important, but not an exclusive, criterion for categorization.

Inspection of Table II reveals a suggestive trend in the degree of object-oriented play among species of primates: the complexity of interactions with objects increases as the more highly evolved forms of primate life are considered. Despite all the complexities of prosimian locomotor and social play, object-oriented play is conspicuously absent in wild populations, and manipulation and exploration of items other than food or the locomotor substrate are rare (Doyle, 1974). When prosimians are placed in a laboratory environment, however, exploratory and manipulative responses are sometimes directed toward inanimate objects in the colony enclosure. Whether the

instances of object-directed behavior in captive prosimians
are the result of a lack of social or locomotor play, increased exposure to novel objects, or some other inferred variable (see Section III), the disparity between laboratory and
field reports has led Jolly (1966) to conclude that interaction with objects in these species is largely an artifact of
captivity.

As we move up the scale of primate species, we find that
New World species, both in captive *and* in free-ranging situations, explore and manipulate objects in the environment, but
in the free-ranging populations, these activities do not develop into play with objects. Most reports on free-ranging
species are careful to convey the idea that behavior directed
toward objects is limited to exploration and manipulation.
Only two reports suggest that captive New World primates play
with objects. Both Moynihan (1976) and Oppenheimer (personal
communication) describe instances of captive *Cebus* (sp. *apella*
and *capucinus*) tossing a handkerchief or a ball around the
cage with their tails.

The reason that prosimians and New World primates do not
show frequent play with objects is not directly testable;
however, several explanations can be suggested. First, Jolly
(1964) has found that the responses of prosimians to nonfood
objects are remarkably similar to their responses to food
items. Those species which prey on insects are likely to have
long attention spans and exhibit little manipulation of objects, while those that are herbivorous are likely to show
short bouts of attention and greater amounts of manipulation.
The propensity to manipulate objects is, therefore, shown to
be directly related to the food habits of a prosimian species,
and the best performers in Jolly's test of manipulative abilities in prosimians were captive, herbivorous species.

Second, the morphology of the hand may predispose a species to use or not use an object in play. The hand and tail
morphology of New World primates is adapted for locomotion in
an arboreal environment; the hands of howlers (*Alouatta*) and
titis, sakis, and uakaris (*Pithecia*) have the first two fin-
gers of each hand opposed to the remaining three, and capu-
chins (*Cebus*) and spider and woolly monkeys (*Ateles*) have
functional prehensile tails (Moynihan, 1976). These adaptations facilitate grasping branches, but preclude the fine
manipulations required for object-play, alteration of the environment, and tool-use. It is perhaps these adaptations to
arboreal life and ecologically-determined responses to objects,
and not a lack of ability to understand object relations, that
preclude the extensive use of objects by infants and juveniles
and the widespread functional expression of tool-use in adult
forms.

In 1968, Jay wrote that "[in monkey play] little, if any,

attention is directed to play with objects or to manipulation
of many objects in the environment" (1968:502). Since then,
many reports of Old World monkey play have appeared which indicate that object-play is an obvious part of the ontogeny of an
individual, a part that is critical for gaining valuable environmental information. It is in the Old World primates that
one first sees the extensive use of objects as a focus of and
means of supporting play (see Table II).

It is in the lesser and greater ape species that we see
the most widespread and complex incorporation of objects into
play sequences. The chimpanzee's ability to manipulate and
play with objects has been well known since Köhler's classic
work in the early part of this century, and recent reports on
the more elusive ape species indicate that object relations
account for a great proportion of play in the young of those
species. The siamang provides a useful example of the importance of object-play in early development. Siamang groups
typically consist of several adult members and several offspring that are a few years apart in age, so infants and juveniles are not found in the large peer groups that are usually
seen in other ape species and the majority of Old World species. Chivers (1974) observed a great amount of solitary object-play in young siamang and proposed that this type of activity effectively substitutes for peer group play.

Of possible interest with regard to the categories of play
is a description of the ontogeny of object-play in a single
individual. In the Old World species for which descriptions
of this process are available, the development of this type of
play is sufficiently similar among species to warrant a general descriptive ontogeny. Menzel (1966) and Fedigan (1972)
provide latitudinal descriptions for the ontogeny of object-
play in Japanese macaques (*Macaca fuscata*) and vervets (*Cercopithecus aethiops*), and Hall and DeVore (1965), Meier and
Devanney (1974), and Jay (1965) utilize a longitudinal strage-
gy for baboons (*Papio anubis*), rhesus monkeys (*M. mulatta*),
and common langurs (*Presbytis entellus*).

A general ontogeny of object-play among Old World primates
begins with the fact that an infant rarely leaves its mother
during the first few months of life, only occasionally going a
few feet away from the mother to investigate food sources that
the mother is eating. Locomotor play, usually solitary, is
observed in two- to four-month-old infants, along with inter-
personal exploratory play directed toward other members of the
age-class. At four to six months of age, when the infant begins to eat solid food, exploration of environmental objects
with the hand and mouth is seen. This is also the stage at
which the infant begins moving farther away from its mother,
in the process probably acquiring information regarding the
environment that may contribute to its survival. From six

months to two years of age, the infant progresses from solitary self-amusement with objects to the beginnings of more complicated play, often supplementing social forms of play by the inclusion of objects. Finally, solitary play with objects often seems to serve as an invitation for social object- or nonobject-play, object-play perhaps serving a metacommunicative function in the sense of van Hooff (1972) and Andrew (1963).

Evidently, familiarity and play with an object as an infant or juvenile is requisite for the expression of object manipulation as an adult. In Goodall's (1968) monograph on the chimpanzees of the Gombe Stream Reserve, she reports that although young chimpanzees sleep with their mothers until about four years of age, they also construct rudimentary sleeping nests during the day, often in a playful context. This "nest-making" play appears in chimpanzees at about one year of age. As they repeat these constructions over a period of two to three years, they become so proficient at the task that they are capable of building their own nests after leaving their mothers. That the expression of adequate nest-making is dependent upon play with nest materials as a juvenile is suggested by comparisons between feral-born and laboratory-born chimpanzees (Bernstein, 1962). When a wild-born chimpanzee, taken from its natural habitat at a few years of age, is given nest materials, it will promptly build a nest, whereas laboratory-raised chimpanzees separated from their mothers a few days after birth will not build nests.

A situation somewhat similar to nest-making may be seen in the development of "termite fishing" among chimpanzees. No infants under two years of age were observed poking twigs into termite holes, yet from about nine months of age, young chimps were seen to exhibit a great interest in the "fishing" activity of their mothers. A one- to two-year-old infant will grasp twigs and prepare them for use as fishing tools (e.g., stripping the bark, modifying its length). At two to three years of age, young chimpanzees are making crude attempts at fishing, often using inappropriate materials. By four years of age, chimpanzees utilize the complete adult technique and are often successful. Again, we see a pattern of infant attention to objects, crude attempts at using and manipulating them, and, finally, after prolonged experience and practice, the appearance of the functional adult form of object use. No direct analogy of the fishing behavior has been studied under controlled laboratory conditions, but it is likely that experience with twigs and grasses as a young animal is essential for the expression of adult "termite-fishing".

Experimentally controlling young animals' access to objects in the environment is a popular and useful technique employed to examine the effects of interaction with objects in a playful situation during youth on object-use later in life.

Menzel (1964) raised young chimpanzees in a gradient of environments, including environmental restriction, access to a
social partner but not to objects, and a complex environment
that encouraged interactions with objects, conspecifics, and
human observers. When tested at two years of age for responsiveness to objects, only the animals raised in the complex
environment showed no fear reactions and approached the test
object promptly. After prolonged exposure to the objects, the
restriction-reared animals' fear of objects eventually sub-
sided and they exhibited tentative exploratory grasps and
sniffs. As play with the test objects developed, differences
in the nature and complexity of behavior among the groups were
seen. The restriction-reared animals showed more self-directed behavior with the objects, including self-stimulation of
body surfaces and simple prehension of objects. The animals
raised in the complex environments quickly passed through this
stage of self-directed behavior and progressed to more complex
forms of object-play, such as using the test object to explore
and manipulate other objects in the environment. In a similar
experiment utilizing older chimpanzees (six to eight years of
age) in a food-getting task, Menzel *et al.* (1970) found that
restriction-reared animals showed poorer performances in using
an object to obtain food than wild-born animals raised in complex environments. Using more abstract tasks such as puzzle
completion and size discrimination, Rumbaugh, Riesen and
Wright (1972) obtained additional evidence that differential
early experience with objects in the environment yields different levels of responsiveness to objects in both juveniles
and adults.

To us, the literature on nonhuman primates suggests that
(1) play is directed toward inanimate objects sufficiently
often to qualify this behavior as a specialty of the Order,
(2) object-play is a category of sufficient standing to de-
serve special analysis, (3) there are clear phyletic differences in the quality and quantity of play among nonhuman primates, (4) captivity has dramatic effects on play, (5) ontogenetic reports on single individuals of different species
show similar patterns of development of play, and (6) there is
an evident relationship between the quality of the early environment in regard to the possibility of play and the quality
of adult behavior.

V. PLAY IN HUMAN CHILDREN

The most remarkable aspect of children's play is its range
and diversity. This is perhaps most apparent, at least in
modern Western cultures, in the proliferation of the manufac-

tured play-things -- dolls, wagons, jump ropes, and other toys
-- that are so frequently incorporated into children's play.
Children are not only recipients of culturally provided play-
things, however, but are also creators of instruments of play.
Goodman (1971) has pointed out that in cultures with relatively few commercially made playthings, there may still be a
great deal of "play technology", i.e., know-how in fashioning
play materials from resources available in the environment.
This technology is transmitted by both adults and older children. As examples, Goodman cites observations of the play objects of Chama and Guaraya children in the Bolivian jungles.
Chama children have been observed spending considerable time
creatively shaping toys, using leaves for umbrellas and toy
cups, reeds for whistles, and tree juices to blow bubbles.
Guaraya children have been known to manufacture elaborate and
inventive toys, including kites, stilts, marbles and balls.
Although play technology appears to have been hindered, or at
least modified, by modern urbanization, the spontaneous fashioning of toys continues, as is evident in Uttely's (1976) reflections on her rural chilhood: flower spikes were used as
soldiers, rushes served as whips, and burrs were fashioned into baskets and nests.

There has been surprisingly little study of children's
creative fashioning of objects for play, although this form of
behavior in nonhuman primates has received widespread attention. We found but one cursory report, that of a study of
Hungarian children's making of toys for invented play sequences (Vince-Bakonyi, 1969). In contrast, the ontogeny of children's use of objects in play, particularly during the sensory-motor period of infancy, has been explored thoroughly,
perhaps because chronological age is a convenient variable.
This section concentrates on the ontogenetic literature in
evaluating the adequacy of the categories formulated in Section III. No attempt is made to review all reported instances
of object-play in children; they are too numerous. The aim,
as in Section IV, is to highlight examples of play that reveal
important classificatory distinctions; it should be noted,
however, that these examples are largely of middle-class North
American infants. Although cultural and individual differences in object-play are obviously important variables, they
raise issues unrelated to our purposes here. Similarly, an
evaluation of factors that influence object-play (e.g., the
amount and types of objects available, dimensions of familiarity and complexity, cultural or parental influences, etc.) is
beyond the scope of this section.

Examples of each category of object-play -- solitary, social, and with other objects -- are in evidence within the
first twelve to fifteen months of infancy. Generally speaking, the behavior of infants proceeds from sensory-motor

activity for its own sake to play involving relationships
(e.g., cause-effect, means-ends) between actions and objects,
social and solitary, animate and inanimate. As we shall see
in the next section, the same ontogenetic pattern appears among
all primates. Piaget (1951, 1952) described fundamental steps
with regard to inanimate objects. In the first few months of
life, objects appear to have no instrumental role in the activities of infants. Behavior is initially reflexive and develops into the repetition of actions for their own sake ("primary circular reactions"). At this age, play appears to consist of the simple pleasure of functioning (e.g., sucking to
suck, looking to look). Such play is clearly nonobject. A
change occurs at three or four months when infants become interested in altering their environment. This is first evident
in simple attempts to prolong interesting sights, as in kick-
ing to make a mobile move ("secondary circular reactions").
Shortly thereafter, infants begin applying more general procedures (e.g., shaking, banging, rubbing) to many objects.
Such reactions might be considered as the first instances of
Object-play/Solitary; however, as yet the focus is on feedback
from the object itself, not on some use to which the object is
put. Piaget argues that "objects" have not yet been elaborated. This occurs in the next stage, at about one year of age,
in which "means" and "ends" are clearly differentiated, ensur-
ing "a new setting into relationship of objects among themselves" (1952:263). Within this stage, Piaget describes cases
of "discovering new means by experimentation" (e.g., using a
stick to obtain an object), cases which clearly fall into the
category Object-play/Object.

Fenson *et al.* (1976) have described a similar progression
in the development of manipulative play. They found that the
play of seven-month-olds was characterized by banging, mouthing, and exploring single objects. Toward the end of the
first year, however, infants began to relate one object to another, either through associations (e.g., cup on saucer) or
groupings (e.g., putting all the cups together). Fenson *et al.*
(1976) pointed out that the infant's attention to these relations coincides with attention to other relations, such as
cause-effect, as is evident in a one-year-old's fascination
with wind-up toys, switches, etc., and means-ends, as is described by Piaget. Similarly, Gessell *et al.* (1940) have described a number of adaptive behaviors emerging in the second
year which involve putting two objects into a constructive relationship with each other. These include building two-block
towers, putting a cube in a cup, and fitting forms in a form
board.

The development of social play is marked by a similar pattern of learning to manage relationships, but this time with
respect to social objects. Development can be seen as pro-

gressing from ignorance of social objects, to contingent interactions (similar to secondary circular reactions), to managing
functional, coordinated social interactions (Mueller and Lucas,
1975). The role of objects in this development has been reviewed recently by Mueller and Vandell (in press). They cite
the observations of Maudry and Nekula (1939), Lichtenberger
(1963), and Mueller and DeStefano (n.d.), who have similarly
noted a shift in peer contacts from a focus on the object ("ob-
ject-centered contacts") to a focus on the peer, with the object used as a means of social interactions (what we have categorized as Object-play/Social). For example, while at first
children frequently come into contact by mutually manipulating
an object, or even by imitating another child's use of an object, in these cases the primary focus is still on the object,
not on the other child. Later, objects are used to elicit another child's attention or interaction and are employed as a
functional part of social games. Such games, including offering and receiving objects and playing ball and peek-a-boo, have
been observed occurring within the first year (Bridges, 1933;
Lichtenberger, 1963; Maudry and Nekula, 1939; Rheingold, Hay
and West, 1976).

Although objects may be employed by infants as a means of
social interaction, they are not necessary for social interaction. Contingent sensory-motor play interactions often occur
between parents and infants without the presence of mediating
objects (see Brazleton, Koslowski and Main, 1974). Mueller and
Vandell (in press) have hypothesized that mediate physical objects may be more instrumental in establishing interactions
between peers than between parents and infants. Nonobject peer
play ranges from simple contingent interactions through run-
and-chase, rough-and-tumble play to complex games of tag or
guessing. Thus, the category of non-object play subsumes many
instances of play with social objects, instances that exhibit
a high degree of ontogenetic sophistication.

Despite the impression of a hierarchy that one may gain
from Table II of Object-play, Object-play/Social and Object-
play/Other, objects do not appear to follow a strict ontogenetic order in children. They appear to involve similar levels
of complexity and to emerge at roughly the same time. Of
course, a child's orientation and experience may lead to an
ordering of skills. Bell (1970) and Decarie (1974), for example, have found that infants tend to be more advanced in
their concept of persons than in their concept of inanimate objects, probably because of the greater affective salience that
other human beings have for children. On the other hand, the
development of peer relations appears in some ways to lag behind that of sensory-motor skills, peers being somewhat more
unreliable respondents than inanimate (or parental) objects
(Mueller and Vandell, in press).

Beyond the sensory-motor period, children's play with objects becomes elaborate. One prominent new element is symbolization. In large part, early symbolization can be viewed as an outgrowth of children's actions involving objects. Piaget (1951) describes make-believe episodes beginning when certain ritual behaviors with respect to one object (sleeping on a pillow) are used with new objects (a piece of cloth or fur). Another early case of symbolic elaboration appears when soft, cuddly objects, which initially were simply a source of contact comfort, begin to be endowed with the qualities of a living companion (Freud, 1965; Singer, 1973). At first, symbolic play occurs in brief, typically solitary episodes, but later it is elaborated, both socially and textually, into "sociodramatic" play, which often incorporates a rich variety of objects (Smilansky, 1968).

In symbolic play, objects could be regarded as "tools" for representation. Yet, unlike tools, as symbols the physical qualities of the objects become increasingly unimportant. Whereas young children appear to rely on concrete substitutes for represented objects, with aging the nature of the symbol becomes increasingly arbitrary, and its concrete representation, unnecessary (Overton and Jackson, 1973; Piaget, 1951). Older children are able to act in a completely imaginary way with no substitute object whatsoever. In this case, the absence of objects in play (non-object play) is a mark of ontogenetic maturity. Purely imaginary play is practically unknown among nonhuman primates, a remarkable exception being the chimpanzee Vicki's fleeting play with an imaginary pull toy (Hayes, 1952). (One worries, however, that the apparent human/nonhuman break may be a creation of our own imagination as human observers.)

In addition to symbolic uses, objects may be incorporated in increasingly complex sensory-motor skills, social games, and constructive activities. In the first case, children's striving for mastery (White, 1959) quickly goes beyond their own bodies to include such objects as swings, jump ropes, and stilts; objects which extend, elaborate, and challenge personal performance. Much of this play appears as Object-play/Solitary, for it is seemingly focused on personal achievement. Nevertheless, the social implications should not be overlooked. Even though the play is solitary in focus, the presence of other children is often facilitating. Mueller and colleagues have noted that the use of a toy by one toddler draws others into "object centered" contacts (see Mueller and Vandell, in press). Aldis (1975) has observed that children's use of slides and swings appears to be stimulated by and largely dependent upon the presence of other children. In these cases, the primary focus of play remains solitary, with the social functions being indirect. In other cases, particularly in competition or showing off, the function of the play becomes explicitly social. Our categories are not well-equipped to separate symbolic from concrete situations.

Games such as marbles and basketball employ objects within highly structured social play. As in the case of symbols, the value of the objects in such cases is essentially determined by their use; in this case, they are used in a social context. Aldis (1975) has noted that in the simple game of competing for an object, the object competed for, be it a stick or a ball, has no great intrinsic value to the player. Its value is determined by the social context of competition. It is easy to imagine a developmental progression of social games from initial "object centered" contacts (where an object is valued because it is desired, used, or occupied by another) toward more complex coordinations of personal actions and objects within social perspectives and goals (Mead, 1934; Piaget, 1948).

Another dimension of children's play is the development of the constructive use of materials (blocks, sand, clay, paint, and so on). Freud (1965) describes an early transition from play with "objects", namely cuddly toys, to play with materials that serve personal activities and fantasies. This appears to be a transition from Object-play/Solitary to the use of objects (materials) in relation to other objects (Object-play/Other objects). Psychoanalytic theorists have generally emphasized the transition of young children's play from play-for-its-own-sake toward the more product-oriented activity of the latency period. Apart from any theoretical perspective, it is evident that children's activities often proceed from non-productive play and imitation toward the assumption of more instrumental, constructive ends. It is, therefore, important that we discriminate the development of serious instrumental "work" from play. In this regard, Bruner (1972) has argued that play provides opportunities to experiment with new combinations of behavior, opportunities which he believes were critical to the evolution of tool-using. Presumably, play provides an opportunity to experiment with new combinations of or relations between objects. Certainly, Bruner's example of the solving of Köhler's classic rake-and-food problem is as much a matter of combining objects as it is of combining behaviors. Sylva, Bruner and Genova (1976) have more recently presented some empirical support for a relation between play and preschoolers' solution to a rake problem.

In summary, the play of children is readily assimilated within the categories of object play. The categories appear particularly useful in describing the focus of play and the development of object relations in infancy. Several problems are encountered, however. While the categories are partially adequate reflections of ontogeny, the richness of older children's play leads to some difficulties. For instance, our

scheme cannot adequately deal with the developing symbolic aspect of play. In addition, the category of Object-play/Soli-
tary has been stretched: while it has been primarily regarded
as indicating play that focuses on what the individual can do
(effectance/mastery), it has also been used to include instances of play that merely occur in isolation. In either case, it
is rare that any play with objects does not involve at least
some relationships to other objects; hence, the dividing line
between Object-play/Solitary and Object-play/Other objects is
not always clear.

The long developmental course of children's play indicates
that no category is restricted in its ontogenetic sophistication. Many instances of play, exhibiting all degrees of complexity, are independent of mediation by physical objects.
When objects are used in play, they appear to have central
roles in the development of peer relations and object relations, and, in human beings, in the emergence of symbolism.

VI. THE INSTRUCTIVE CASE OF *Macaca fuscata* AND THEIR SWINGS

In this section, we present observations on the Bucknell
colony of Japanese macaques *(M. fuscata)* which illustrate several of the categories outlined in Section III. The animals in
the first phase of our study were seven members of the colony,
three older females, two younger adult females, an adult male,
and a juvenile male, living in a 12 x 12 m enclosure that fur-
nished trees and other objects designed to encourage naturalistic behavior [see Candland *et al.* (1972) for a detailed description of the field cage]. Figure 1 shows the ages and
genealogy of the colony.

A. *Original Fashioning of the Swing*

During late November, 1975, a steel bar 2 m by 7 mm in dia-
meter fell from the superstructure of the enclosure into the
cage. After the bar had remained on the ground for a period of
several weeks, we noted that it had been bent into the shape of
S, inserted into the metal framework within the enclosure, and
was being used as a swing by the juvenile male. After several
days of observation, we decided to remove the bar and replace
it with similar objects in a partially counterbalanced design
in order to assess the generality of manipulating an object into swings and to measure the development of the use of the object for play. The new objects were a similar unbent bar, a
chain with hooks attached to both ends, and a chain with unbent
bars on either end, called the bar-chain. On each day of

observation, one of the objects was presented for an hour in the morning and another for an hour in the afternoon. Observers recorded the following behavior: time and frequency of grasping and biting, dragging (bipedally or quadrupedally), climbing (with forearms or quadrupedally), inserting the object, and using it as a swing. Presentation of the bar was first alternated with the chain for 25 trials, after which the chain was alternated with the bar-chain for 25 trials.

The juvenile male *(Ag)* was by far the most frequent user of the swing objects, followed by a young adult female *(Af)*. It was also the juvenile male who fashioned the two objects (bar and bar-chain) which required modification to be used as swings. This modification was accomplished by wedging the object on a tree branch or in the wire mesh of the enclosure and then bouncing or tugging on the object (see Figure 2). Once a bend had been made in either end, the object was inserted into the wire mesh on the roof or side of the enclosure and then used as a swing.[5]

Figure 3 describes the development of play with the different swing objects for the juvenile male and young adult female. Several points should be considered: first and most obvious, the juvenile male played with the swing objects far more than did the young adult female. That the objects were modified and used as swings by the juvenile and the youngest adult member of the colony suggests that swinging, like innovative behaviors among *M. fuscata,* is likely to be "invented" or "discovered" and propagated by immature animals (Kawai, 1965).

Second, the juvenile male became more proficient at modifying objects (i.e., learning to fashion swings) as he gained experience with unmodified objects. The frequency of playful interactions, which was low during the first ten sessions with the bar, is probably related inversely to the length of time required to successfully modify the object for use as a swing. When the bar-chain, which also required modification, was

[5]*We did not, unfortunately, witness the initial fashioning of the fallen bar and, hence, can only speculate about the origin of swing fashioning behavior. Because the juvenile male performed all of the modifications of the introduced swing objects required for use as swings, and the method of modification described above resulted in swings similar to the original discovery, we cautiously infer that the juvenile male was the fortuitous originator of swing fashioning and swinging behavior. He was also the greatest user of swings, and similar work on tool-using in primates by Beck (1973, 1976) indicates that the original tool discoverer also shows the highest incidence of tool behavior.*

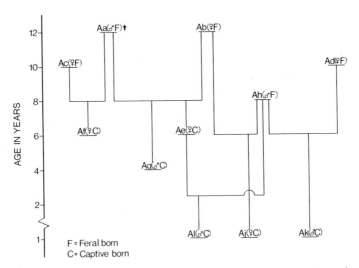

FIGURE 1. Age, sex, origin (feral- or captive-born), and family relationships of the Bucknell colony of M. fuscata.

introduced in session 26, the preparation of the object for use as a swing was almost immediate, this being reflected in the initial high frequencies of play with the bar-chain.

Third, after the period of acquisition of the swinging behavior, both the juvenile male and the young adult female exhibited a decrease in the amount of play with the swing objects. We choose to call this phenomenon "adaptation" of play with objects, since the behavior follows the established principles of adaptation to a novel stimulus. Table III lists the trials that showed the greatest positive slope (acquisition) and the greatest negative slope (adaptation) for swinging and grasping over the course of the study. Both grasping and swinging show slopes significantly different from zero, suggesting acquisition and adaptation of the behavior. Other behaviors show a slope different from zero, reach asymptote promptly, and remain at asymptote during the course of the sessions.

For time spent swinging, acquisition occurs quickly by the fifth trial. Adaptation is also rapid, occurring by the eleventh trial. The peaks of acquisition and adaptation are most rapid for the bar-chain, next for the chain, and slowest for the bar. For many patterns of behavior, the period of most rapid adaptation follows promptly the period of rapid acquisition. This suggests that however creative the act of swing fashioning may be, measurements of behavior show the known principles of acquisition and adaptation.

FIGURE 2. Juvenile male modifying an introduced swing object (bar). The object has been rendered stationary and weight is being applied on a bend in the bar.

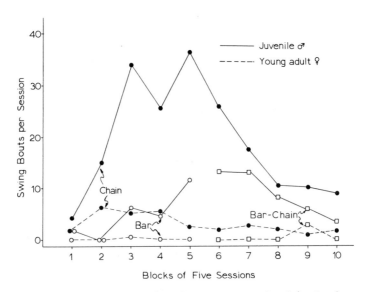

FIGURE 3. Number of swing bouts on each object observed during each session (one hour) for the juvenile male (Ag) *and the young adult female* (Af).

By examining Figure 3, it is seen that the acquisition and adaptation of object-play responses were not specific to each object; swinging generally increased over the first 25 sessions, then showed a marked decrease in the following 25 sessions.

On very few occasions did more than one animal manipulate or play with the swing objects simultaneously. The animal who was playing with the swing at a given time assumed sole ownership of it and did not allow others to play with it (this was especially so if the juvenile male possessed the swing object). Possession of the swing often prompted threats and chases from another animal.

Fashioning an environmental object for the apparent purpose of play, and then using it in play, qualifies as an example of object-play in a solitary situation. Some may construe this fashioning behavior as an example of tool-use or tool-making, and were it not for Alcock's definition constraining our classification of animal-object relations, this activity could be regarded as tool-use for the purpose of play. As the classification stands, however, this instance of play with an object remains an example, albeit a more creative one than most, of the category Object-play/Solitary.

TABLE III. Trials, Slopes, and p Values[a]

	Bar (25)				Chain (50)				Bar-Chain (25)			
	TS	RS	TG	RG	TS	RS	TG	RG	TS	RS	TG	RG
Trials showing maximum acquisition	1-4	17-20	18-21	12-15	2-5	11-14	7-10	15-18	1-4	1-4	3-6	3-6
Trials showing maximum adaptation	4-7	22-25	21-24	15-18	8-11	14-17	24-27	31-34	6-9	6-9	6-9	11-14
Slope acquisition	.72	1.45	2.91	2.52	2.12	5.55	5.63	6.95	.46	3.25	3.35	3.70
Slope adaptation	-.72	-1.60	-1.40	-2.94	-4.44	-7.65	-5.93	-6.60	-.38	-2.15	-7.90	-6.05
p Acquisition	.30	.30	.05	.10	.05	.30	.05	.20	.10	.02	.20	.20
p Adaptation	.30	.30	< 1	.05	.20	.20	.30	.05	.02	.10	.05	.10

[a]TS = Time swinging;
RS = Rate swinging;
TG = Time grasping;
RG = Rate grasping.

B. *Infant Acquisition of Swinging*

During the observation of swing fashioning, three infants
were born into the colony. We wished to see if and when the
infants would exhibit swinging behavior and what categories of
play they would exhibit. Because the fashioning and prehension
of swing objects require strength greater than that of infants,
we suspended the originally modified swing from an eyehook
placed in the branch of a tree. Observers recorded sequences
and durations of play interactions with the swing for one-hour
sessions, five days a week. By placing the swing inside the
enclosure fifteen minutes prior to the beginning of an observation session and taking it out immediately afterward, we were
able to observe all the responses of colony members to the
swing, the ontogeny of swinging behavior, and the dissemination
of the play pattern. Our comments here concern the ontogeny of
the swinging response and its relation to our model of play
categories.

The infants exhibited interest in the swing objects at an
early age (two to three months). This interest usually took
the form of watching the swing area while the juvenile male was
playing with the swing. It is likely that the sight of the
juvenile male swinging back and forth and the auditory sensa-
tions resulting from the swing being moved about the enclosure
attracted the infants' attention. Another aspect of this visual orientation was the infants' tendency to observe the juvenile male play while they were in close proximity to the area
around the swing.

At about six months of age, an important change in behavior toward the swing was noted: physical contact was made with
the swing. These initial contact sequences were characterized
by four discrete stages: (1) lengthy visual orientation to
the swing area, (2) hesitant approaches to the swing, (3) rapid
grasping, mouthing, or sniffing of the swing object, and finally (4) a retreat to the mother or a nearby social (nonobject)
play group. This exploratory sequence was repeated several
times daily by each of the infants. As time passed, the approaches to the swing object became less tentative and the exploratory behaviors more intense and prolonged.

The criterion for differentiation of exploration and play
with objects developed in Section IV was used to establish the
onset of play with the swing by the infants. Specifically,
when an active motor response such as climbing, swinging, or
jumping on and off the swing object was observed in the exploratory sequences, play with the swing was said to have occurred.
Age at onset of play was variable, the female *(Aj)* beginning at
eight months of age, one male *(Al)* at eight and one-half
months, and the other male *(Ak)* at eleven months. An infant's
age at onset of play with the swing was related to its mother's

relative status in the colony hierarchy: the more dominant
the mother, the earlier that play with the swing was observed
in her infant.

We noted in the general descriptive ontogeny of play with
objects that solitary object play can occur as early as two
months of age and usually is expressed by six months, yet in
this case, object-play, as we define it, did not appear until
eight to eleven months of age. This discrepancy is probably
due to several factors, including the complexity of the swinging response compared to other forms of object-play. Infants
in our colony have been observed tossing sticks and stones
about the enclosure, playing "keep-away" with small, shiny objects, and engaging in other forms of object-play as early as
seven weeks of age, but the swing presents a much larger visual stimulus, moves back and forth in the wind or after an animal has played with it, and is presumably more novel than
other objects in the enclosure. These stimulus factors, plus
the fact that the swing was introduced for only an hour each
day, may account for the longer period of habituation necessary with this object. Also, play with the swing requires a
certain level of motor proficiency, including the ability to
jump accurately, to climb up and down, and to maintain balance.
These requisites for swinging are not necessary for more se-
date forms of object play, such as tossing sticks, but swinging is a very active task, requiring strength and coordination
lacking in very young infants.

Early interactions with the swing were always solitary and
of very short duration, and although they contained active
play responses, the sequences were usually dominated by exploratory and manipulative behaviors. Figure 4a presents a
frequency flow diagram of typical play sequences derived from
data on Al during the early period of his interactions with
the swing. Note that he directs only three behaviors toward
the swing, two of which are exploratory. There is the suggestion of a stereotyped pattern, one behavior following another
a great proportion of the time, and there is much repetition
and redundancy in the play sequence. During this period of
early interactions with the swing, the frequency of play bouts
within individuals was highly variable from day to day, and
the juvenile male easily displaced the infants when he wanted
to play with the swing.

Figure 4b provides a similar flow diagram derived from
data on the same animal taken a month and a half later. There
was an obvious change in the complexity of play sequences, as
shown by the ways in which the animal combines more responses
in different ways and exhibits fewer repetitions of responses
and sequences of responses. Inherent in such variable sequences is a potential for developing "new games" on a daily
basis, with the concomitant changes in novelty-induced arousal.

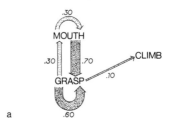

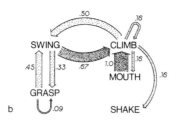

	0 -30%
	30-60%
	60-above
100% = 1.0 cm.	

FIGURE 4.   (a) Frequency flow diagram of play sequences
directed at the swing by Al, an eight and one-half month old
male. Key provides proportions of sequences of behavior.
(b) Frequency flow diagram of play with the swing for Al, now
10 months old.

After all infants had reached the stage of interacting
playfully with the swing, another shift in the nature of object-play was observed: social play became the dominant form
of infant object play. Typical forms of rough-and-tumble play,
grasping and tugging responses, and other types of peer play
were displayed while the infants were also climbing and swinging on the object (see Figure 5). To develop a measure of the
amount of social object-play in relation to solitary play
bouts, the following calculation was performed:

$$\text{Index of Social Object-Play} = \frac{\text{Number of Social Object-Play Bouts}}{\dfrac{\sum \text{Number of Bouts per Individual}}{\text{Total Individuals Playing}}}$$

The result is a scale ranging from zero to one. A value of
zero indicates that no social object-play bouts were observed

FIGURE 5. *Object-play/Social. All three infants playing on the swing or in close proximity to the swing.*

in a given day, while a value of 1 means that all object-play
bouts in a given day were of a social nature. Figure 6 presents these values summed over each week of an eleven-week
period during which the transition from solitary to social object-play was most striking. During the first three weeks,
solitary swing interactions were the sole form of object-play
exhibited by the infants, but during the following weeks there
was a marked increase in the proportion of social play bouts.
The function eventually asymptotes at around 0.75, showing
that an average of three out of every four play bouts observed
involved more than one animal.

Infrequent observations of social cooperation between infants during play bouts were made. We have recorded on film
an incident of one infant sitting on the swing as another runs
back and forth on the ground, pulling the swing as he moves,
providing motion for the infant on the swing (see also Figure 7). These observations of a high degree of social play
and of instances of cooperation between animals are in contrast to the absence of such activities during swing fashioning and during initial infant play when the animals were
younger.

We have seen (Sections III and IV) that our model effectively separates primate taxonomic groups on the basis of the
extent to which inanimate environmental objects are incorporated into play sequences. An additional utility of the categories appears to be that they describe the ontogeny of object-play sequences among individuals.

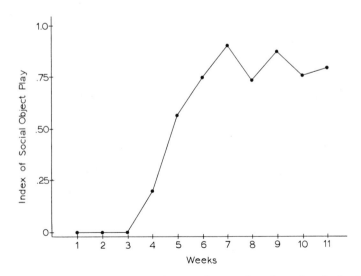

FIGURE 6.   Index of social object-play by the infants for
an 11-week period. See text for derivation and discussion.

FIGURE 7.   Object-play/Social.   The infant on top provides
motion for the swing by pulling and pushing with his hind legs
while the infant on the bottom rides the swing.

Initial responses to the swing were primarily grasps, mouthings, and sniffs, all behaviors of the category Exploration and Manipulation. Our initial inclusion of exploration and manipulation appears to be justified, not merely because of the difficulties encountered in distinguishing these activities from playful ones, but also because both appear to be essential to the process of adaptation to novel objects and to the fullest expression of object-play. The exploratory and manipulative activities were followed by individual play interactions on the swing, such interactions becoming more complex and increasingly variable, yet still solitary. Play activities culminated with behavior categorized as Object-play/Social dominating the infants' interactions with swing objects.

The ontogenetic changes in object-play follow the increasing complexity of animal-object play relations seen in the primate species review of Section IV. That the model suggests the direction of the ontogeny of object-play, as well as confirms phyletic relationships, lends support to the accuracy and utility of our provisional categories.

VII. CONCLUSION

In this chapter, we have set out categories of play and examined their credibility by examining the characteristics and ontogeny of an unusual form of play by *M. fuscata*. Insofar as this singular form of play has generality toward other forms of play, it appears that the categories are worthy separators.

While categorization is not explanation, it is certainly fundamental to it. The categorical system must now await the application of past and present theories of play to its demands as a test of their generality. If valid, the categories should not be unduly bent by the application of viewpoints described elsewhere in this volume, for such application should clarify troublesome places in these theories while supplying a common language to describe and evaluate observations.

ACKNOWLEDGMENTS

We thank Paul Weldon, Gregory Lorinc, Solveigh Scott, Caren Rosenthal, Michael Pereira, Jon Jensen, Mark Puccetti, and Kevin O'Connor for their work on this project. Weldon and Lorinc were also involved in an earlier analysis of the swinging behavior described in Section VI. Mary Nornhold clarified the text.

REFERENCES

Alcock, J. The evolution of the use of tools by feeding animals. *Evolution 26,* 464-473 (1972).

Aldis, O. "Play Fighting". Academic Press, New York (1975).

Altmann, S. A. Field observations on a howling monkey society. *J. Mammal. 40,* 317-330 (1959).

Andrew, R. J. The origin and evolution of the cells and facial expressions of the primates. *Behaviour 20,* 1-109 (1963).

Aries, P. "Centuries of Childhood". Knopf, New York (1962).

Baldwin, J. D. The ontogeny of social behavior of squirrel monkeys *(Saimiri sciureus)* in a seminatural environment. *Folia primatol. 11,* 35-79 (1969).

Beck, B. Cooperative tool use by captive hamadryas baboons. *Science 182,* 594-597 (1973).

_____ Tool use by captive pigtailed macaques. *Primates 17,* 301-310 (1976).

Bell, S. M. The development of the concept of object as related to infant-mother attachment. *Child Develop. 41,* 291-311 (1970).

Bernstein, I. S. Response to nesting materials of wild born and captive born chimpanzees. *Anim. Behav. 10,* 1-6 (1962).

_____ Activity patterns in a gelada monkey group. *Folia primatol. 23,* 50-71 (1975).

Bishop, A. Control of the hand in lower primates. *Ann. N. Y. Acad. Sci. 102,* 316-337 (1962).

Blurton-Jones, N. G. An ethological analysis of some aspects of social behaviour of children in nursery schools, *in* "Primate Ethology" (D. Morris, ed.), pp. 437-463. Aldine, Chicago (1967).

Brazelton, T. B., Koslowski, B., and Main, M. The origins of reciprocity: the early mother-infant interaction, *in* "The Effect of the Infant on Its Caregiver" (M. Lewis, and L. A. Rosenblum, eds.), pp. 49-76. John Wiley and Sons, New York (1974).

Bridges, K. M. R. A study of social development in early infancy. *Child Develop. 4,* 36-49 (1933).

Bruner, J. Nature and uses of immaturity. *Amer. Psychol. 27,* 687-708 (1972).

Bruner, J., Jolly, A., and Sylva, K. (eds.). "Play: Its Role in Development and Evolution". Basic Books, New York (1976).

Butler, R. A. Investigative behavior, *in* "Behavior of Nonhuman Primates, Vol. 2" (A. M. Schrier, H. F. Harlow, and F. Stollnitz, eds.), pp. 463-493. Academic Press, New York (1965).

Candland, D. K., Geiger, W., and Bell, J.   A field cage for
    Old World primates designed to encourage natural behavior.
    Primates 13, 315-322 (1972).
Chivers, D. J.   The siamang in Malaya: a field study of a
    primate in a tropical rain forest.   *Contrib. Primatol. 4*,
    1-335 (1974).
Decarie, T. G.   "Intelligence and Affectivity in Early Child-
    hood: An Experimental Study of Jean Piaget's Object Con-
    cept and Object Relations".   International Universities
    Press, New York (1974).
Dolhinow, P. J., and Bishop, N.   The development of motor
    skills and social relationships among primates through
    play. *Minn. Symp. Child Psychol. 4*, 141-198 (1970).
Doyle, G. A.   Behavior of prosimians, *in* "Behavior of Nonhuman
    Primates, Vol. 5" (A. M. Schrier, and F. Stollnitz, eds.),
    pp. 155-353.   Academic Press, New York (1974).
Doyle, G. S., Andersson, A., and Bearder, S. K.   Maternal be-
    haviour in the lesser bushbaby *(Galago senegalensis moholi)*
    under semi-natural conditions. *Folia primatol. 11*, 215-
    238 (1969).
Eaton, G.   Snowball construction by a feral troop of Japanese
    macaques *(Macaca fuscata)* living under seminatural con-
    ditions. *Primates 13*, 411-414 (1972).
Fedigan, L.   Social and solitary play in a colony of vervet
    monkeys *(Cercopithecus aethiops)*. *Primates 13*, 347-364
    (1972).
Fenson, L., Kagan, J., Kearsley, R. B., and Zalazo, P. R.   The
    developmental progression of manipulative play in the
    first two years. *Child Develop. 47*, 232-236 (1976).
Freud, A.   "Normality and Pathology in Childhood".   Inter-
    national Universities Press, New York (1965).
Gessell, A., Halverson, J. W., Thompson, H., Ilg, F. L., Cast-
    ner, B. M., Ames, L. B., and Amatruda, C. S.   "The First
    Five Years of Life".   Harper, New York (1940).
Goodman, M. E.   "The Culture of Childhood: Child's Eye View
    of Culture and Childhood".   Teacher's College Press, New
    York (1971).
Groos, K.   "The Play of Animals".   Chapman and Hall, London
    (1898).
_____   "The Play of Man".   D. Appleton and Co., New York
    (1901).
Hall, G. S.   "Youth: Its Education, Regimen, and Hygiene".
    Appleton, New York (1906).
Hall, K. R. L.   Tool-using performances as indicators of be-
    havioral adaptability. *Curr. Anthropol. 4*, 479-494 (1963).
Hall, K. R. L., and DeVore, I.   Baboon social behavior, *in*
    "Primate Behavior: Field Studies of Monkeys and Apes" (I.
    DeVore, ed.), pp. 53-110.   Holt, Rinehart and Winston,
    New York (1965).

Hayes, C. "The Ape in Our House". Excerpt (1952) reprinted in "Play: Its Role in Development and Evolution" (J. Bruner, A. Jolly, and K. Sylva, eds.), pp. 534-536. Basic Books, New York (1976).

Hutt, C. Exploration and play in children, *in* "Child's Play" (R. E. Herron, and B. Sutton-Smith, eds.), pp. 231-251. John Wiley and Sons, New York (1971).

Jay, P. C. The common langur of north India, *in* "Primate Behavior: Field Studies of Monkeys and Apes" (I. DeVore, ed.), pp. 197-249. Holt, Rinehart and Winston, New York (1965).

_____ (ed.). "Primates: Studies in Adaptation and Variability". Holt, Rinehart and Winston, New York (1968).

Jewell, P. A., and Loizos, C. (eds.). "Play, Exploration, and Territory in Animals". Academic Press, New York (1966).

Jolly, A. Prosimians' manipulation of simple object problems. *Anim. Behav. 12,* 560-570 (1964).

_____ "Lemur Behavior: A Madagascar Field Study". University of Chicago Press, Chicago (1966).

Jung, C. G. "Collected Works". Pantheon Press, New York (1953).

Kawai, M. Newly-acquired precultural behavior of the natural troop of Japanese monkeys on Koshima islet. *Primates 6,* 1-30 (1965).

Köhler, W. "The Mentality of Apes". Harcourt Brace, New York (1925).

Lichtenberger, W. Mit nenschlickes verhalten eines Zwellingspaares in sein in erst en lebensjahren. No. 48, Ernst Reinhardt Verlag, Munich. Reviewed by Mueller and Vandell (in press) (1963).

MacKinnon, J. The behaviour and ecology of wild orangutans *(Pongo pygmaeus). Anim. Behav. 22,* 3-74 (1974).

Maudry, M., and Nekula, M. Social relations between children of the same age during the first two years of life. *J. Genet. Psychol. 54,* 193-215 (1939).

McCall, R. B. Exploratory manipulation and play in the human infant. *Monogr. Soc. Res. Child Develop. 39,* 1-88 (1974).

Mead, J. H. "Mind, Self, and Society". University of Chicago Press, Chicago (1934).

Meier, G. W., and Devanney, V. D. The ontogeny of play within a society: preliminary analysis. *Amer. Zool. 14,* 289-294 (1974).

Menzel, E. W. Patterns of responsiveness in chimpanzees reared through infancy under conditions of environmental restriction. *Psychol. Forsch. 27,* 337-365 (1964).

_____ Responsiveness to objects in free-ranging Japanese monkeys. *Behaviour 26,* 130-150 (1966).

_____ Spontaneous invention of ladders in a group of young chimpanzees. *Folia primatol. 17,* 87-106 (1972).

_____ Further observations on the use of ladders in a group
of young chimpanzees. *Folia primatol. 19,* 450-457 (1973).

Menzel, E. W., Davenport, R. K., and Rogers, C. M. The development of tool using in wild-born and restriction-reared
chimpanzees. *Folia primatol. 12,* 273-283 (1970).

_____ Protocultural aspects of chimpanzees' responsiveness to
novel objects. *Folia primatol. 17,* 161-170 (1972).

Millar, S. "The Psychology of Play". Pelican, London (1968).

Moynihan, M. "The New World Primates". Princeton University
Press, Princeton (1976).

Mueller, E., and Lucas, T. A. Developmental analysis of peer
interactions among toddlers, *in* "Friendship and Peer Relations" (M. Lewis, and L. A. Rosenblum, eds.), pp. 223-258.
John Wiley and Sons, New York (1975).

Mueller, E., and DeStefano, C. Sources of toddler's peer interaction in a playgroup setting. Unpublished manuscript
(n.d.).

Mueller, E., and Vandell, D. Infant-infant interaction, *in*
"Handbook of Infant Development" (J. Osotasky, ed.).
John Wiley and Sons, New York (in press).

Overton, N., and Jackson, J. The representation of imagined
objects in action sequences: a developmental study.
Child Develop. 44, 309-314 (1973).

Piaget, J. "The Moral Judgment of the Child". Free Press,
Glencoe, Ill. (1948).

_____ "Play, Dreams, and Imitation in Childhood". Routledge
and Kegan Paul, London (1951).

_____ "The Origins of Intelligence in Children". International Universities Press, New York (1952).

Poirier, F. E., and Smith, E. O. The crab-eating macaques
(Macaca fascicularis) of Anguar Island, Palua, Micronesia.
Folia primatol. 22, 258-306 (1974).

Reynolds, V., and Reynolds, F. Chimpanzees of the Budongo
Forest, *in* "Primate Behavior: Field Studies of Monkeys and
Apes" (I. DeVore, ed.), pp. 368-424. Holt, Rinehart and
Winston, New York (1965).

Rheingold, H. L., Hay, D. F., and West, M. J. Sharing in the
second year of life. *Child Develop. 47,* 1148-1158 (1976).

Rosenblum, L. A., and Cooper, R. W. (eds.). "The Squirrel
Monkey". Academic Press, New York (1968).

Rothe, J. Some remarks on the spontaneous use of the hand in
the common marmoset *(Callithrix jacchus). Proc. Int.
Congr. Primatol., Zurich, 3,* 136-141 (1970).

Rumbaugh, D. M., Riesen, A. H., and Wright, S. C. Creative
responsiveness to objects: a report of a pilot study with
young apes. *Folia primatol. 17,* 397-403 (1972).

Schaller, G. B. "The Mountain Gorilla: Ecology and Behavior". University of Chicago Press, Chicago (1963).

_____ The behavior of the mountain gorilla, *in* "Primate

Behavior: Field Studies of Monkeys and Apes" (I. DeVore, ed.), pp. 324-367. Holt, Rinehart and Winston, New York (1965).

Singer, J. L. "The Child's World of Make-Believe: Experimental Studies in Imaginative Play". Academic Press, New York (1973).

Smilansky, S. "The Effects of Sociodramatic Play on Disadvantaged Preschool Children". John Wiley and Sons, New York (1968).

Spencer, H. "Principles of Psychology". Appleton, New York (1873).

Stevenson, M. F., and Poole, T. B. An ethogram of the common marmoset *(Callithrix jacchus jacchus)*: general behavioral repertoire. *Anim. Behav. 24*, 428-451 (1976).

Sylva, K., Bruner, J. S., and Genova, P. The role of play in the problem solving of children 3-5 years old, *in* "Play: Its Role in Development and Evolution" (J. S. Bruner, A. Jolly, and K. Sylva, eds.), pp. 244-257. Basic Books, New York (1976).

Uttely, A. Field toys, *in* "Play: Its Role in Development and Evolution" (J. S. Bruner, A. Jolly, and K. Sylva, eds.), pp. 199-201. Basic Books, New York (1976). Original source: "Country Hand". Faber and Faber, New York (1943).

van Hooff, J. A. R. A. M. A comparative approach to the phylogeny of laughter and smiling, *in* "Nonverbal Communication" (R. A. Hinde, ed.), pp. 209-241. Cambridge University Press, Cambridge (1972).

van Lawick-Goodall, J. The behaviour of free-living chimpanzees in the Gombe Stream Reserve. *Anim. Behav. Monogr. 1*, 161-311 (1968).

_____ Tool-using in primates and other vertebrates, *in* "Advances in the Study of Behavior, Vol. 1" (D. Lehrman, R. Hinde, and E. Shaw, eds.), pp. 195-249. Academic Press, New York (1970).

Veblen, T. "The Theory of the Leisure Class". Modern Library Inc., New York (1934).

Vince-Bakonyi, A. Self-made toys in children's games. *Int. J. Early Childhood 1*, 15-19 (1969).

Warren, J. M. Tool use in mammals, *in* "Evolution of Brain and Behavior in Vertebrates" (R. B. Masterson, M. E. Bitterman, G. B. G. Campbell, and N. Hotton, eds.), pp. 407-424. Lawrence Erlbaum Associates, Hillside, New Jersey (1976).

Weisler, A., and McCall, R. B. Exploration and play: résumé and redirection. *Amer. Psychol. 31*, 492-508 (1976).

Welker, W. I. An analysis of exploratory and play behavior in animals, *in* "Functions of Varied Experience" (D. W. Fiske, and S. R. Maddi, eds.), pp. 278-325. Dorsey Press, Homewood, Illinois (1961).

_____ Ontogeny of play and exploratory behaviors: a defini-

tion of problems and a search for new conceptual solutions, *in* "The Ontogeny of Vertebrate Behavior" (H. Moltz,
ed.), pp. 171-228. Academic Press, New York (1971).
White, R. Motivation reconsidered: the concept of competence.
Psychol. Rev. 66, 297-333 (1959).

A LONGITUDINAL STUDY OF SOCIAL PLAY IN
SYNANON/PEER-REARED CHILDREN[1]

Karen H. Hamer
Elizabeth Missakian

Synanon Research Institute
Marshall, California

*Over 3,000 bouts of social play were recorded during 300
hours of observation of free play in 36 Synanon/peer-reared
children (9-45 months of age). The total number of play bouts
comprised 7,857 individual play interactions among the children. Social play is defined as any non-maintenance, non-
agonistic interaction in which the participation of one child
in an activity engages the interest and participation of another child or children. Information was compiled on gender
differences, age differences, group size, gender combinations,
use of roles, use of equipment, and prevalence of type of play.*

I. INTRODUCTION

In the past 5-10 years, there has been an awakening interest in play behavior, its form and function, in both human and
nonhuman research. This short review will not deal with the
ever-increasing volume of work on animal play, despite the obvious parallels with various forms of human play. The reader
is referred to excellent reviews by Loizos (1967), Millar
(1972) and Baldwin and Baldwin (1977). In the realm of child
development, much pioneering work was done by Parten (1932)
and Bridges (1933), who established some concepts and assumptions concerning young children's social behavior -- concepts

[1]*This research was supported by Synanon Foundation, Inc.,
and a grant from the National Institute of Mental Health.*

297

which continue to be given credence. Parten (1932) lists and
defines six categories of social participation in the preschool
aged child, which are: unoccupied behavior, onlooker, solitary
independent, parallel activity, associative play, and organized supplementary play. With the exception of the last two
categories, these are self-explanatory. Associative play is
group play in which there is overt recognition by group members
of their common activity, interest, and personal association,
while supplementary cooperative play is organized purposefully
for some goal; it has definite membership and directing members. Bridges (1933) writes that young children (9-10 months
of age), if given time to associate, have the beginnings of
play that is "becoming socialized and mutually enjoyable"
(p. 43).

In the years following the 1930's, little work was done on
the social relations of groups of young children. Children's
social play was not again seriously pursued until the late
1960's. During the intervening years, there was much descriptive work on the individual child's play, i.e., motor development, stages and psychology of play in the individual. Weisler
and McCall (1976) provide an excellent review of this work.
Piaget (1951) lists some commonly-held characteristics of play:
autoletic, spontaneous, pleasurable, lack of organization, and
occurring when there is freedom from conflict. These are
strikingly similar to many characteristics often given for nonhuman primate play. Piaget further states that these are inadequate explanations of play behavior and concludes that play
is the assimilation of reality to the ego. Smith and Connally
(1972) categorized play in the following manner: (a) activity,
looking on, nothing; (b) moving or stationary; (c) self, parallel, or group; (d) toy or no toy. All of these categories
utilize nearly exactly Parten's definitions. Blurton Jones
(1972), in his definition of social behaviors, defined the
category "play with" in the following way: "implies they (the
children) are using the same play materials and perhaps ex-
changing and discussing them" (p. 106). He also states that
"the overt behavior involved in dramatic play divides into
three types: rough and tumble play, social exchange of objects, and imitations of adult activities" (p. 21).

There is general agreement that all play components appear
in other behavioral categories, i.e., smile, talk to, reach
out, touch, run, chase, etc. This seems to be a major difficulty in defining play, since it has no specific components
(with the possible exception of the play face in nonhuman primates) and, therefore, can only be identified relative to the
context in which it appears, by the reactions of the other
participants and, finally, by default since most behavior
which seems to serve no discernible primary purpose is often
labelled play (Millar, 1962).

The evolutionary significance of play in nonhuman mammals is increasingly being emphasized (Aldis, 1975; Beach, 1945; Poirier and Smith, 1974), stimulating the emergence of new concepts of the function of social play in children. Bruner (1972) hypothesized that the ability to play is developed as a central theme of evolution and may be a major harbinger of the development of language and symbolic behavior, that play is an opportunity for trying out the reordering of behavioral sequences with little risk, and occurs in a context that reduces excessive drive and frustration.

Bronson (1975), in her study of pairs of unacquainted toddlers, concluded that even though initiation of the peer encounter is not frequent, a young child's rate of engagement remains consistent and is predictive of nursery school peer behavior. Grief (1976), in her study of 3-5-year-old children, was struck by the large amount of play interaction. Although younger children interacted less frequently, it was only because they lacked the necessary skills and not because they lacked interest in social engagement. In this analysis of sex role playing, Grief found that an average of 44% of the children's speaking was imaginative and that there was no sex difference in the amount of participation in imaginative play. Leacock (1971) described children's play as conscious experimentation with environments and abilities. Research on nonhuman primates (Jay, 1965; Mason, 1965) concurs with Fortes' (1938) speculation that social play is a rehearsal of social living bereft of penalties.

For all the seductiveness of these theories concerning children's play and its function, the truth seems to be that very little field work has been done on children in situations which are natural and uninterrupted. Playing with unfamiliar children in unfamiliar or semi-familiar settings with strong valenced adults present is not a good indicator of either the capabilities or inclinations of the young. Anecdotal evidence/ observations of the anthropologists are at least as valid and certainly more interesting than the drawing of generalizations about the phylogenetic significance of children's play from 15 minute play bouts of pairs of unfamiliar children.

This paper presents the parameters of social play in a naturally-occurring stable group of young children living in peer groups whose daily behaviors affect one another. The questions asked are: (1) Which children interact with each other and with what frequency? (2) What behaviors comprise play? (3) What is the most consistent group size? (4) What are the effects of age and gender on social play? (5) What roles do children assume during social play?

II. SYNANON FOUNDATION, INC.

A. *The Synanon Community*

Beginning with approximately 20 individuals in 1958, Synanon Foundation, Inc., has grown in 17 years into a community
housing over 1,300 men, women and children. For a detailed
description of the history and current status of Synanon, see
Yablonsky (1965), Endore (1968) and Simon (1974). Synanon currently has communities in Santa Monica, San Francisco, Badger,
and Marshall, California. Synanon International is the international branch of the Foundation, with chartered organizations
in Germany, Malasia and Manila.

During the early years of its history, Synanon was best
known for its work in the rehabilitation of drug addicts and
alcoholics. Since 1958, over 13,000 people have come to Synanon for help and, as a result, over 2,200 Synanon-like drug
programs have been established throughout the United States.
The primary business of Synanon continues to be "the people
business". In addition to this work, Synanon is involved in
establishing innovative methods of communal/peer child-rearing,
health care delivery systems, community design, and general
life style. One of Synanon's best known efforts in education
is The School.

B. *The Synanon School*

The Synanon School began in July, 1966. Prior to that
time, children of Synanon residents either lived outside of the
community or lived with their parents and attended public
schools. In 1966, a decision was made to begin 24-hour-a-day
child care for the community. The School has grown from 11
children in 1966 to over 240 children (from birth to 18 years
of age) at the onset of this study.

Our observations of the children living in Synanon focused
on the group living in the Lower School in Marshall, California. This particular program is the home of children ranging
from 6 months to 4 years of age. The Lower School has a capacity of 30 children, housed in one main building (487.5 m^2
floor space). Within this building are two workshops/class-
rooms, two sleeping rooms, a kitchen/dining room, two toilet/
changing rooms, and an adult work and storage area. The building was designed to staff specifications for convenient, efficient, and flexible child care. The children's sleeping areas
are divided into rooms which contain 3-4 individuals each.
Each child's bed space is his/her inviolate area and contains
particular possessions and is decorated with pictures of

family, friends and favorite activities. The workshops are 9.14 m x 15.24 m, well-lit, attractively decorated rooms. Usually, one area in the room is partitioned off as a classroom for formal adult-directed activities. The rest of the room contains age-appropriate areas and equipment for dramatic play, block play, reading, puzzles, and large muscle activities (ramps, platforms, swings, ladders, waterbed, and ropes). These areas are used for free play. The configuration of the rooms, the equipment and the decor are changed frequently.

The outdoor space of the Lower School is divided into three open yards, totalling over 1,579 m^2. The "grass yard" is adjacent to the Infant Workshop and is, in fact, grassy. It contains toys, arts/crafts areas, swings, and a portable swimming pool. The "large yard" is gravel and contains a trampoline, sandboxes, bikes and other rolling stock, jungle gym, hide boxes, platforms, and covered porches. The "upper yard" is a tricycle yard. During free play, the children have access to all equipment and play areas in the yards. The children are encouraged to play out-of-doors and are provided with all-weather gear to facilitate this play. Generally, they have free access to either indoor or outdoor space. In addition to the enriched space described above, children are regularly (daily) taken on field trips both within and outside the Synanon community.

The staff/child ratio in the Lower School is one adult to four children. The staff are all members of the Synanon community. The School provides in-service training, as well as employing a number of accredited teachers who are also community members. The School is approved by the State of California.

III.   METHODS

A.   Subjects

During the period of study, the population of the Lower School fluctuated between 23-30 children. Also, during that period, children moved into and out of this particular program. The total number of children observed was 36, ranging in age from 9-47 months. The mean age of children observed during 1975 ranged from 11.2 - 45.5 months. The mean age in 1976 was from 19 - 46.5 months. Of the 36 children, 15 were girls and 21 were boys. In 1975, there were 11 girls with a mean age of 24.75 months and 19 boys with a mean age of 27.64 months. In 1976, there were 15 girls with a mean age of 32.97 months and 16 boys with a mean age of 34.39 months. In 1976, the boys were on the average 6.75 months older than in 1975 and the

girls were 8.61 months older. See Table I for demographic
information on the children of Synanon.

B. Observation Schedule and Recording Methods

    The daily activity schedule of the children generally con-
sisted of periods of free indoor and outdoor play, workshops,
field trips, and rest, in addition to routine care such as
eating, bathing and toileting. Observations were made during
both indoor and outdoor free play situations. The amount of
time spent in free play varied with the age of the child and
the season of the year, but averaged 3-4 hours daily. Generally, 2-3 hours of observations were collected each day. Records were maintained on the adult's presence in the play area.
If an adult disrupted the children's activity for more than
several minutes, observations were discontinued. The observer
sat or stood in the play area in full view of the children.
Occasionally, the observer would move closer to the children
and only in cases of extreme aggression or danger interfered
in their activities. Therefore, the children seldom turned to
her for assistance or entertainment.
    During the first year of study, data were collected by con-
stant scanning of the observation area and verbally describing
the interactions into a tape recorder. Whenever possible, the
verbal conversation of the children playing was recorded. The
tapes were transcribed and analyzed for play interactions. In
the second year, a form was developed on which all essential
elements of the interaction were recorded. Each interaction
was tabulated according to date, time, number of players, type
of play, role of each player, sex composition of the play
group, location of the interaction, type of vocalization, and
role assumption.
    A total of 300 hours of observation were recorded. Two
hundred hours in 1975 were recorded and 100 hours in 1976.
Records were maintained of each child's observation time. The
range was from 11 to 84 hours per child in 1975 and from 12 to
60 hours per child in 1976. Some children moved in and out of
the program and some children were much less observable than
others. No attempt was made to equalize the amount of time
each child was observed. Instead, all results were weighted
for observation time per child. This is a preliminary study of
social play in young children and, for that reason, the observer was interested in overall patterns of play rather than
questions of individual differences and variability. If a
child was not seen by the observer, he/she was also not seen
by peers and was not as potentially influential in social play.
There were several other reasons for variability in participa-

TABLE I. Name, Gender, Birthdate, Number of Interactions,
Number of Interactions per Hour and Interaction Rank
for 1975 and 1976

			Interactions		No. of Interactions per Hour		Interaction Rank	
Name	Gender	Birth	1975	1976	1975	1976	1975	1976
Br	male	5/71	113	---	2.65	---	13.0	---
Em	female	6/71	61	---	1.55	---	22.0	---
Ao	male	12/71	299	---	4.06	---	2.0	---
Dm	male	1/72	209	---	2.98	---	10.5	---
Jd	male	3/72	221	54	3.46	4.60	7.0	4.0
Ja	male	3/72	124	---	2.39	---	16.0	---
Dl	female	5/72	273	81	3.84	4.97	3.0	3.0
Ch	male	6/72	142	111	2.69	4.02	12.0	7.0
No	male	7/72	---	81	---	3.85	---	8.0
Ar	male	7/72	264	94	3.65	2.89	5.0	17.0
Ey	female	8/72	---	26	---	2.12	---	27.0
Sm	male	8/72	234	120	3.48	4.25	6.0	6.0
Ma	male	10/72	231	86	3.71	2.82	4.0	19.0
Aa	female	1/73	180	189	2.44	3.71	15.0	10.0
Za	male	1/73	371	179	4.96	5.55	1.0	1.0
Ru	female	1/73	---	151	---	2.87	---	18.0
Si	female	1/73	213	212	3.15	4.40	9.0	5.0
Jo	male	3/73	81	153	2.52	3.14	14.0	13.0
Jf	female	3/73	159	179	2.35	3.74	17.0	9.0
Lh	female	3/73	101	192	1.45	3.25	23.0	12.0
Jr	male	5/73	250	153	3.22	3.07	8.0	15.0
Is	male	6/73	251	257	2.98	5.01	10.5	2.0
Ln	female	7/73	85	102	2.12	2.29	18.0	23.5
Ca	female	8/73	67	83	1.97	2.46	19.0	21.0
Fr	female	8/73	---	103	---	3.12	---	14.0
Mx	male	8/73	113	73	1.35	1.52	24.0	30.0
Al	female	10/73	139	75	1.96	1.75	20.0	29.0
Ek	female	1/74	---	93	---	2.92	---	16.0
Me	female	2/74	---	74	---	3.44	---	11.0
Js	male	2/74	23	33	1.18	2.32	28.0	22.0
Di	male	3/74	23	111	1.32	2.29	26.0	23.5
Mi	male	5/74	26	101	1.28	2.15	27.0	26.0
Cs	female	6/74	26	124	1.95	2.68	21.0	20.0
Ga	male	7/74	17	108	1.12	2.19	29.0	25.0
Cr	male	8/74	3	41	0.20	0.85	30.0	31.0
Ra	female	8/74	27	84	1.34	2.00	25.0	28.0

tion in the peer group: visiting with friends or family, toi-
let training (a time-consuming activity), illness and field
trips.

Definitions of social play, types of play, and roles of
play evolved from observation of these children from 1972-1975.
Social play in children is defined as any non-maintenance, non-
agonistic activity in free time/play in which the participation
of one child in an activity engages interest and participation
of another child or children. All social play interactions in
which these children engaged were either motor/skill play or
symbolic-imaginative/dramatic play. *Skill play* is play in
which the players and their extensions, i.e., toys and equipment, are used in the real or actual sense. That is, a slide
is a slide -- it is not a ski slope and the child using that
slide is, himself, not a bear or a world famous skier. Conversely, *dramatic play* is play in which the players and their
extensions are, by some wonderful mental deception, that which
they are not -- Sam is a troll or the waterbed is an ocean.
Specific motor patterns and behaviors often overlapped in these
categories and only in the context of the total situations
could an interaction be defined as dramatic or skill play. See
Table II for a listing of various behaviors seen in each category of social play.

Participants in the play episodes were designated as initiators, imitators, cooperators, resistors, or observers. An
initiator is a child who begins a play sequence. An *imitator*
is a child who, seeing or hearing the play activity, then attempts to follow or repeat the activity. A *cooperator* is a
child who, when invited to play by another, joins in the play
sequence; a cooperator might also spontaneously join in the
play. A *resistor* is a child who, when invited to play, rejects
or ignores the invitation or attempts to stop or divert the
play. An *observer* is a child whose attention is obviously concentrated on the play activity, but who does not join in.

IV. RESULTS

Each social play bout involved two or more children; therefore, the total number of individual interactions is greater
than the total number of play bouts. The total number of individual interactions is referred to as "interaction frequency".
Table III contains information on the number of hours observation, number of social play bouts, and interaction frequency
for 1975 and 1976. Results for each year are presented sepa-
rately, since the age/sex composition of the groups varied between the two years. Play frequencies per child were weighted
for amount of time (in hours) each child was observed.

TABLE II. Behavior Categories of Dramatic and Skill Play[a]

Dramatic Play		Skill Play	
squeal	watch	lay down	giggle
cover eyes	spin swing	cover	look out
bring	lay down	jog	wear hat
quack	swim	ride trike	call
look in	run	chase	teeter totter
"meow"	sit	read	invite
park	offer	peek	extend hand
drink	take	bounce	ride on/in
give medicine	be monster	bend	vocalize
play music	be an animal	look through legs	smile
look in mouth, ear,	scream	drag	smack
or eye	invite	roll in	drop
take bath	feed	jump	hit
diaper	eat	tickle	be hit
stir fire	climb	load truck	empty
shoot	bark	build	wind tire swing
stomp	step in	spin	ring round rosey
touch w/object	flee	reach for	stack
toot	chase	squeeze object	climb
knock over objects	yell at	stand on	poke
manipulate toys	line up chairs	hide behind	grab
verbalize	spread blanket	grin	laugh
offer	rev bikes	lay in swing	slide

Continued on following page

Dramatic Play

Skill Play

shake hand	"sleep on"	wipe	carry
reprimand doll	put on clothes	balance	sit on
crawl as animal	take temperature	line up	point
push object	be the patient	growl	wrap in curtain
pour	hug	lead	blow whistle
receive	stand in	follow	sing
ride, drive toy	"water" horse	stand in bucket	push object
fix	bite w/puppel	carry bucket	walk
lick	serve	sit under platform	hold hands
	be served		

[a]Categories of dramatic and skill play are not exhaustive and are not meaningful unless viewed in the context of the entire social play situation.

TABLE III. *Number of Hours Observation, Number of
Play Bouts, and Total Play Interaction
Frequency for 1975 and 1976*

Year	No. of Hours Observation	No. of Play Bouts	Total Play Interaction Frequency
1975	200	1,700	4,354
1976	100	1,341	3,503

A. 1975

A total of 1,700 play bouts were observed during 1975.
The average frequency of play bouts per hour of observation
was 8.50. The average number of children seen playing per
hour was 21.32. The average number of play bouts per hour per
child was 2.45.

1. *Gender.* In the 1,700 play bouts observed in 1975, the
play interaction frequency of the 30 children (19 males and 11
females) was 4,354. Table IV contains the frequency of play
for males and females and Table V contains the gender x age x
interactions per hour. The females played 1,319 times and the
males played 3,035 times. These frequencies were adjusted for
the number of each gender and amount of time observed. The
females played 2.19 times per hour and the males played 2.59
times per hour. Tables VI and VII contain summary tables for
analyses of variance on play data for 1975 and reveal no gender differences in play frequency. Data from Table VIII provide the basis for these analyses.

TABLE IV. *Gender, Interaction Frequency, Number of Play
Interactions, and Percentage of Total Play Interactions
for 1975 and 1976*

Year	Frequency	% of Total Interactions	No. of Children	% of Total
1975				
Male	1,319	30.3	11	36.7
Female	3,035	69.7	19	63.3
1976				
Male	1,757	50.1	16	52.0
Female	1,746	49.8	15	48.0

TABLE V. *Frequency of Interactions per Hour and Number of Children per Group for Males and Females in Six Various Age Groups for 1975 and 1976*

| | Females | | | Males | |
Age Group	No. of Interactions per Hour	No. of Females per Age Group	Age Group	No. of Interactions per Hour	No. of Males per Age Group
1975					
6-12 months	0.43	1	6-12 months	0.45	5
13-18 months	2.15	4	13-18 months	1.39	6
19-24 months	1.76	5	19-24 months	1.64	5
25-30 months	2.01	5	25-30 months	3.02	7
31-36 months	3.58	3	31-36 months	3.70	8
36+ months	3.26	2	36+ months	4.33	8
1976					
13-18 months	1.32	1	13-18 months	2.65	1
19-24 months	2.78	2	19-24 months	1.92	5
25-30 months	2.63	4	25-30 months	1.61	4
31-36 months	2.72	8	31-36 months	3.39	4
36+ months	3.65	7	36+ months	3.84	9

TABLE VI. Summary Table Analysis of Variance:
Gender x Age x Type of Play for 1975

	SS	df	MS	F
Between	34.34	85		
Age	6.72	3	2.24	6.59[a]
Gender	.66	1	.66	1.94
Age x Gender	.66	3	.22	.65
Subj. Within	26.30	78	.34	
Within	29.10	86		
Type of Play	.04	1	.04	.14
Age x Type of Play	7.21	3	2.40	8.57[a]
Gender x Type of Play	.02	1	.02	.07
Age x Gender x Type of Play	.02	3	.006	
C x Subj. Within	21.81	78	.28	

[a] $p \leq .05$

TABLE VII. Summary Table Analysis of Variance:
Gender x Age x Role of Play for 1975

	SS	df	MS	F
Between	15.44	105		
Age	3.66	4	.915	8.32[a]
Gender	.09	1	.09	.82
Age x Gender	.67	4	.17	1.55
Subj. Within	11.02	96	.11	
Within	155.33	424		
Roles	17.92	4	4.48	13.58[a]
Age x Roles	6.49	16	.41	1.24
Gender x Roles	2.59	4	.65	1.97
Age x Gender x Roles	1.72	16	.11	.33
Subj. Within	126.61	380	.33	

[a] $p \leq .05$

TABLE VIII. *Frequency of Interactions, Number of Subjects, Number of Hours of Observation, and Number of Interactions per Hour for 1975 by Age Groups*

Age Group	Total Interactions	No. of Subjects	Hours of Observation	Interactions per Hour
6-12 mos.	19	6	49.95	0.38
13-18 mos.	234	10	141.71	1.65
19-24 mos.	622	10	326.23	1.91
25-34 mos.	1,027	12	366.55	2.80
31-36 mos.	1,138	11	333.78	3.41
36+ mos.	1,313	10	362.10	3.63

2. *Age*. The data were examined for age group information. Table VIII contains interaction frequency, number of hours observation, interactions per hour, and number of children per age group. The range of interactions was 19 (0.38 per hour) for the 6-12 month olds to 1,313 (3.63 per hour) for the 36+ month olds. Tables VI and VII above demonstrate that age is a significant factor in social play. Table IX contains information on the statistical differences between the various age groups for play frequency. The youngest children (6-12 months) play significantly less than all other age groups. The 13-18 month olds and the 19-24 month old children played significantly less than the 36+ month old children ($p \leq .05$). The three oldest age groups spanning 25-36+ months of age were not significantly different from one another in play frequency.

TABLE IX. *Test on Differences Between All Pairs of Means of the Six Age Groups Play Bout Frequency, 1975*

Age (Mos.)	Mean Interaction Freq./Hr.	6-12 $\bar{X}=.38$	13-18 1.65	19-24 1.91	25-30 2.80	31-36 3.41	36+ 3.63
6-12	.38	--	1.27[a]	1.53[a]	2.42[a]	3.03[a]	3.25[a]
13-18	1.65		--	.26	1.15	1.76	1.98[a]
19-24	1.91			--	.89	1.50	1.72[a]
25-30	2.80				--	.61	.83
31-36	3.41					--	.22
36+	3.63						--

[a] $p \leq .05$

3. *Types of Play*. Of the 1,700 play bouts observed in
1975, 716 (42%) were categorized as dramatic play and 984 (58%)
as skill play. Table VI above contains information on the
analysis of variance for age x gender x types of play. Total
frequencies for each type of play were not significantly different. However, the interaction between age and type of play
was significant and indicated that as children grew older,
they engaged in significantly more dramatic play. This is primarily related to the acquisition of language, since (as can
be seen in Table II), many of the behavior categories are overlapping and can only be determined in the context indicated by
language.

4. *Roles*. Of the five roles designated, initiator accounted for 1,533 interactions, cooperator for 1,781, imitator
for 684, resistor for 245, and observer for 111 interactions.
The total frequency for roles was 4,354, since each participant in the 1,700 bouts occupied one of the designated roles.
Table VII above contains the summary table for the age x gender x role analysis of variance. This analysis revealed that
roles was a significant factor. Initiator and cooperator
roles occurred significantly more frequently than other roles.

5. *Group Size*. Table X presents information on group
size for all observed play bouts. Of the total 1,700 bouts
observed in 1975, 64% were two-player groups, 22% were three-
player, 9% were four-player, 3% were five-player, and 1% were
interactions with more than five players.
Equipment/toys were used during 1,250 bouts (73% of total).
Fewer than 6% of the play bouts graded or escalated into aggression.

B. *1976*

A total of 1,341 play bouts were observed during 1976.
The mean number of play bouts per hour was 13.41. The mean
number of children seen playing per hour was 35.03. The mean
number of play bouts per hour per child was 3.1.

1. *Gender*. In the 1,341 bouts, the interaction frequency
of the 31 children (16 boys, 15 girls) was 3,503. Table IV
presents the interaction frequency of play for males and females. Table V shows the gender x age group x interactions
per hour for observations in 1976. The females played 1,746
times and the males played 1,757 times. When these frequencies were adjusted for the number of each gender and the
amount of observation time, the females played 3.04 times per
hour and the males played 3.11 times per hour. Tables XI and

TABLE X.   *Frequency of Play Bouts for Various*
Group Sizes for 1975 and 1976

	Interactions		% of Total	
Gender x Group Size	1975	1976	1975	1976
male/male x 2	559	259	32.80	19.00
female/female x 2	101	259	5.90	19.00
female/male x 2	430	330	25.20	25.00
male/male x 3	125	59	7.30	4.40
female/female x 3	2	49	.12	3.65
female/male x 3	255	202	15.00	15.00
male/male x 4	44	14	2.50	1.00
female/female x 4	0	14	.00	1.00
female/male x 4	114	71	6.70	5.00
male/male x 5	5	3	.29	.22
female/female x 5	0	1	.00	.07
female/male x 5	48	50	2.80	3.73
greater than 5	17	30	1.00	2.22

XII contain summary tables of analyses of variance for play
data for 1976. They reveal no gender differences in play frequencies.

2. *Age*. The data were examined for age group information. Table XIII contains play frequency, number of hours observation, interactions per hour and number of children per
age group. The range of interactions was 59 (1.7 per hour)
for the 12-18 months old to 1,669 (3.74 per hour) for the 36+
months old. Tables XI and XII demonstrate that as children
grow older, they engage in more social play. Table XIII shows
that there were no statistical differences between the various
age groups for interaction frequency for 1976.

3. *Types of Play*. Of the 1,341 bouts, 55% were categorized as dramatic play and 45% as skill play. Table XI is the
summary table for the analysis of variance for age x gender x
types of play. Total frequencies for each type of play were
not significantly different. However, they are an approximate
reversal of the 1975 data. Once again, the interaction between age and type of play was significant, indicating that
older children (with language skills) engage in more dramatic
play than younger children.

4. *Roles*. Of the five roles designated, initiator accounted for 1,182 interactions, cooperator for 1,644, imitator

TABLE XI. Summary Table Analysis of Variance:
Gender x Age x Type of Play for 1976

	SS	df	MS	F
Between	39.53	87		
Age	4.76	4	1.19	2.77[a]
Gender	.10	1	.10	.23
Age x Gender	1.34	4	.34	.79
Subj. Within	33.33	78	.43	
Within	221.18	88		
Type of Play	0.00	1	0.00	0.00
Age x Type of Play	8.95	4	2.24	.82
Gender x Type of Play	.08	1	.08	.63
Age x Gender x Type	.36	4	.09	.03
C x Subj. Within	211.79	78	2.72	

[a] $p \leq .05$

TABLE XII. Summary Table Analysis of Variance:
Gender x Age x Role of Play for 1976

	SS	df	MS	F
Between	15.22	87		
Age	1.68	4	.42	2.47[a]
Gender	.07	1	.07	.41
Age x Gender	.36	4	.09	.53
Subj. Within	13.11	78		
Within	160.86	352		
Role	12.71	4	3.16	7.18[a]
Age x Role	1.96	16	.12	.27
Gender x Role	.05	8	.01	.02
Age x Gender x Role	1.79	32	.06	.14
C x Subj. Within	144.30	326	.44	

[a] $p \leq .05$

TABLE XIII. *Test on Differences Between All Pairs of Means of the Six Age Groups' Play Bout Frequency, 1976*

Age (Mos.)	Mean Interaction Freq./Hr.	Age				
		13–18	19–24	25–30	31–36	36+
		$\overline{X} = 1.70$	1.21	2.31	3.02	3.74
13–18	1.70	--	.49	.61	1.32	2.04
19–24	1.21		--	1.10	1.81	2.53
25–30	2.31			--	.71	1.43
31–36	3.02				--	.72
36+	3.74					--

for 415, resistor for 115, and observer for 147 observations. Table XII above is the analysis of variance summary table for the age x gender x role of play. The analysis indicates that in 1976, roles were a statistically significant factor. As in 1975, initiator and cooperator occurred significantly more than other roles.

5. *Group Size.* Table X presents the group size information for play bouts. Sixty-three percent of the play bouts were two-player groups, 23% were three-player, 7% were four-player, 4% were five-player, and 2% were groups with more than five players.

In 1976, additional data were collected on vocal modes and on character assumption by the children.

6. *Vocal Modes.* The categories of vocalization were: no vocalization (20% of the bouts), vocalization only (1%), verbal exchange only (7%), vocalization in conjunction with physical action (12%), verbalization in conjunction with physical action (56%), and unrecorded (5%). Data on categories of vocalization were collected for the 1,341 play bouts.

7. *Character Assumption.* Character assumption was involved in 11.38% of the 3,503 individual interactions. Roles delineated were: family members (5% of total play; 43% of all character assumption), Synanon School staff (0.5% of total play; 4.7% of all character assumption), community person (0.6% of total play; 4.9% of all character assumption), animals (5% of total play; 44.1% of all character assumption), Synanon Elders (0.14% of total play; 1.2% of all character assumption), and inanimate objects (0.14% of total play; 1.2% of all character assumption).

C. *Comparison of 1975 and 1976*

   1. Gender. In 1975, the females played 2.19 times per hour and in 1976, they played 3.04 times per hour. In 1975, the boys played 2.59 times per hour and in 1976, 3.11 times per hour. In neither year of observation was there a gender difference in the play interaction frequency. A 2 x 2 contingency table for gender x year revealed no significant differences.

   2. Age. In 1975, the range of interaction rate was 0.38 for the 6-12 month olds to 3.63 for children over 36 months of age. In 1976, the range was 1.21 for the 19-24 month olds or 3.71 for the children over 36 months of age. There was very little variability in interactions per hour per age group in the two years of observation. See Figure 1 for a comparison.

   3. Types of Play. In 1975, 42% of the play bouts were categorized as dramatic play. In 1976, this percentage increased to 55%. In 1975, 58% of play was skill play; in 1976, 45% of play was skill play. Between the two years of observation, the proportions of dramatic play increased and the proportion of skill play decreased.

   4. Roles of Play. Table XIV contains the comparison between 1975 and 1976 with regard to roles of play. It is apparent that the assumption of the various roles is relatively

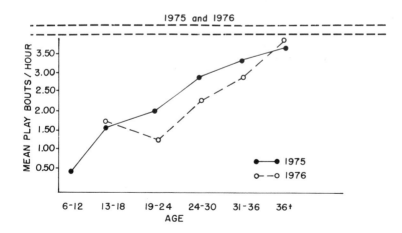

FIGURE 1. Comparison of mean number of play bouts per hour per age group for 1975 and 1976.

consistent during the two years of observation. Initiator and cooperator are the most frequently occurring roles in both years of observation.

5. *Group Size*. The number of players in each play group is consistent over 1975 and 1976. In 1975, 64% of the groups were two-player; in 1976, 63% were two-player. In 1975, 22.4% were three-player groups and in 1976, 23.05% were three-player. In 1975, 9% were four player groups and in 1976, 7% were four-player. In both years of observation, the majority of play occurred in two-player groups.

V. DISCUSSION

Lewis and Rosenblum (1975) speculate, "...the emphasis on adult- as opposed to peer-dominated social experience in our culture is not based on the child's capacities but rather on a very specific ideology" (p. 7). The children living in the Synanon situation represent an opportunity to collect information on the social capacities of the young child. The emphasis on social/peer relations of the Synanon children led us to hypothesize that these children would engage in social play both more frequently and at a younger age than home-reared children. The frequency of social play among Synanon children ranging in age from 12-45 months was in sharp contrast to Bronson's (1975) conclusion, "Given a setting that allows toddlers the freedom of many choices, the data I reported unequivocally show that unacquainted second year age mates characteristically engage in much looking but in little active seeking of social contact with one another" (p. 151). It should not be surprising that children who are acquainted with one another would engage in more social play than unacquainted children. Fur-

TABLE XIV   *Frequency of Various Role Assumptions for 1975 and 1976*

	1975		1976	
Role	*Frequency*	*%*	*Frequency*	*%*
Initiator	1,533	35	1,182	34
Imitator	684	16	415	12
Cooperator	1,781	41	1,644	47
Observer	111	3	147	4
Resistor	245	5	115	3

thermore, it is also no surprise that children reared in 24-hour-a-day peer groups would have frequent and deep social bonds.

We had hypothesized that Synanon-reared children would begin to engage in social play at a younger age than home-reared children. Skill social play was observed frequently among children from 6-12 months of age and imaginative/dramatic play emerged earlier in the Synanon children than children studied by Grief (1976).

The Synanon child-rearing situation is one in which there is minimal sex role stereotyping. Therefore, we expected to find and, in fact, did demonstrate fewer gender differences in play behavior than previously reported. There were no gender differences in play frequency, in types of play, or in role assumption. Males engaged in dramatic/imaginative play as often as females and females were as likely to engage in skill play as were males. Further, our results did not substantiate any evidence that males are the initiators of social play while the females are merely passive participants or observers.

It was interesting to note that dramatic/imaginative play occupied a predominant role in the play of young children reared in the Synanon community. Grief (1976) reported that 44% of 3-5 year olds' articulations were imaginative. Our results demonstrated that for both 1975 and 1976 observations, an average of 48.5% of the social play of Synanon-reared children was imaginative. However, these Synanon children had a mean age of 29.84 months as opposed to the older children in Grief's study. All but the youngest children in our population engaged in some form of dramatic play. Because these Synanon children share many common experiences, they tend to integrate their experiences into their play so that every field trip or extraordinary event becomes dramatized, often with much repetition.

Parten (1932) found that the two-player group was the most prevalent group size in the play of young children. Our observations were consistent with her findings. Parten also found that 68% of the play groups were single sex, while we found that in both years of observation, only 48% of our social play groups were single sex. Once again, play association seemed to be much more a function of age than gender.

The longitudinal approach to the study of child behavior permits analysis of the consistency, not only of an individual child's behavior, but also of patterns of group play behavior. Over the two-year observation period, we were able to determine that the social peer relations of the Synanon children were consistent. Children formed definite friendships which endured over the two-year period.

In summary, the results of our observations indicate that play and peer relations of young children reared in the

Synanon community are quite complex and support Lewis and
Rosenblum's (1975) speculation that, "...adequate opportunities for diverse peer relations early in life may represent a
setting that is quite conducive to substantially normal social
development" (p. 7). We agree wholeheartedly with the above
authors' statement. "We might argue that peers are not just
substitutes for infant-adult relationships, but rather are
just as basic and maybe even older in the phylogenetic sense.
Small families isolated from one another may constitute historically a rather unique and new experience, not only for the
caregivers but for the infants themselves" (p. 7).

Further analysis of our data will focus on individual re-
lationships/friendships of Synanon children, internal group
structure, and individual modes of social play within the
group. We will continue our observations of this particular
grouping of Synanon children to learn the degree to which our
current information is predictive of the social relations of
older children.

REFERENCES

Aldis, O. "Play Fighting". Academic Press, New York (1975).
Baldwin, J. D., and Baldwin, J. I. The role of learning phe-
    nomena in the ontogeny of exploration and play, in "Bio-
    Social Development in Primates" (S. Chevalier-Skolnikoff,
    and F. E. Poirier, eds.), pp. 343-406. Garland Publishing
    Co., New York (1977).
Beach, F. A. Current concepts of play in animals. Amer. Nat.
    79, 523-541 (1945).
Blurton Jones, N. G. Characteristics of ethological studies
    of human behavior, in "Ethological Studies of Child Behav-
    ior" (N. G. Blurton Jones, ed.), pp. 3-39. Cambridge Uni-
    versity Press, London (1972).
Bridges, K. A study of social development in early infancy.
    Child Devel. 4, 35-49 (1933).
Bronson, W. Behavior with age mates during the second year of
    life, in "Friendship and Peer Relations" (M. Lewis, and
    L. A. Rosenblum, eds.), pp. 131-152. John Wiley and Sons,
    New York (1975).
Bruner, J. S. Nature and uses of immaturity. Amer. Psychol.
    27, 687-708 (1972).
Endore, G. "Synanon". Doubleday and Co., Inc., Garden City
    (1968).
Fortes, M. "Social and Psychological Aspects of Education in
    Taleland". Oxford University Press, London (1938).
Grief, E. B. Sex role playing in pre-school children, in
    "Play - Its Role in Development and Evolution" (J. S.

Bruner, A. Jolly, and K. Sylva, eds.), pp. 385-391. Basic
Books, Inc., New York (1976).

Jay, P. C. The common langur of North India, in "Primate Behavior: Field Studies of Monkeys and Apes" (I. DeVore,
ed.), pp. 197-249. Holt, Rinehart and Winston, New York
(1965).

Leacock, E. At play in African villages. Nat. Hist. (Suppl.),
60-65 (1971).

Loizos, C. Play behavior in higher primates: a review, in
"Primate Ethology" (D. Morris, ed.), pp. 176-218. Aldine
Publishing Co., Chicago (1967).

Mason, W. A. The social development of monkeys and apes, in
"Primate Behavior: Field Studies of Monkeys and Apes" (I.
DeVore, ed.), pp. 514-543. Holt, Rinehart and Winston,
New York (1965).

Millar, S. "The Psychology of Play". Penguin Books, Inc.,
Baltimore (1968).

Parten, M. B. Social participation among pre-school children.
J. abnorm. soc. Psychol. 27, 243-269 (1932).

Piaget, J. "Play, Dreams, and Imitation in Childhood". W. W.
Norton and Co., Inc., New York (1951).

Poirier, F. E., and Smith, E. O. Socializing functions of
primate play. Amer. Zool. 14, 275-287 (1974).

Simon, S. I. "The Synanon Game". Unpublished Ph.D. Dissertation. Harvard University, Cambridge, Massachusetts
(1974).

Smith, P. K. and Connolly, K. Patterns of play and social interaction in pre-school children, in "Ethological Studies
of Child Behavior" (N. Blurton Jones, ed.), pp. 65-95.
Cambridge University Press, London (1972).

Weisler, A., and McCall, R. Exploration and play. Amer.
Psychol. 31, 492-508 (1976).

Yablonsky, L. "Synanon: the Tunnel Back". Penguin Books,
Baltimore (1965).

INDEX

A
B
C   8
D   9
E   0
F   1
G   2
H   3
I   4
J   5